Substitutes for Hazardous Chemicals in the Workplace

Per Filskov
Gitte Goldschmidt
Mogens Kragh Hansen
Lena Höglund
Tea Johansen
Christian Libak Pedersen
Lone Wibroe

LEWIS PUBLISHERS

Boca Raton   New York   London   Tokyo

Library of Congress Cataloging-in-Publication Data

Goldschmidt, Gitte.
  Substitutes for hazardous chemicals in the workplace / by Gitte Goldschmidt.
    p.   cm.
  Includes bibliographical references and index.
  ISBN 1-56670-021-3
  1. Hazardous substances--Handbooks, manuals, etc.   2. Substitution (Technology)--
Handbooks, manuals, etc.   3. Industrial safety--Handbooks, manuals, etc.   I. Title.
  T55.3.H3G634 1996
  604.7—dc20   95-49037
                                                                                                             CIP

This book contains information obtained from authentic and highly regarded sources. Reprinted material is quoted with permission, and sources are indicated. A wide variety of references are listed. Reasonable efforts have been made to publish reliable data and information, but the author and the publisher cannot assume responsibility for the validity of all materials or for the consequences of their use.

Neither this book nor any part may be reproduced or transmitted in any form or by any means, electronic or mechanical, including photocopying, microfilming, and recording, or by any information storage or retrieval system, without prior permission in writing from the publisher.

All rights reserved. Authorization to photocopy items for internal or personal use, or the personal or internal use of specific clients, may be granted by CRC Press, Inc., provided that $.50 per page photocopied is paid directly to Copyright Clearance Center, 27 Congress Street, Salem, MA 01970 USA. The fee code for users of the Transactional Reporting Service is ISBN 1-56670-021-3/96/$0.00+$.50. The fee is subject to change without notice. For organizations that have been granted a photocopy license by the CCC, a separate system of payment has been arranged.

CRC Press, Inc.'s consent does not extend to copying for general distribution, for promotion, for creating new works, or for resale. Specific permission must be obtained from CRC Press for such copying.

Direct all inquiries to CRC Press, Inc., 2000 Corporate Blvd., N.W., Boca Raton, Florida 33431.

© 1996 by CRC Press, Inc.
Lewis Publishers is an imprint of CRC Press

No claim to original U.S. Government works
International Standard Book Number 1-56670-021-3
Library of Congress Card Number 95-49037
Printed in the United States of America 1 2 3 4 5 6 7 8 9 0
Printed on acid-free paper

PREFACE

In the Danish Occupational Health Services (OHS) a lot of work concerning substitution of chemicals has been done in the past ten years. The main goal in the work of the OHS has been to reduce the problems with chemicals in the working environment by introducing less hazardous chemicals or by introducing procedures of work where chemicals are not needed.

According to Danish law it is prohibited to use a hazardous chemical if a less hazardous or non-hazardous substitution can be found. It is pronounced clearly by Order No. 540 of 1982 from the Working Ministry.

Moreover, we feel that substitution is a good method to reduce health problems caused by hazardous products. Where devices such as ventilation or exhaust hoods can reduce the problems, substitution simply eliminates the problem causers.

To elaborate the use of substitution even more, it is important to use the obtained results as a platform—the results being good or bad experiences. A lot of experiences of substitution were located in and around 130 OHS's in Denmark. But these experiences were out of reach for working environment professionals outside the single OHS. Therefore, a group of OHS professionals in 1987 set out to collect this knowledge and transform it into a tool that could be used by all OHS professionals in Denmark and also other professionals attached to the work with the chemical working environment, e.g., in The Danish Working Environment Service. The work resulted in a book, which was published in Denmark in 1989 and later translated into German and published in Germany in 1992.

The book is presented in two main parts. In the first part a suggestion for the chronology of a substitution is presented, including different practical hints. A range of tools for evaluating the health risks of chemicals is also presented.

The second part is an adaptation of an inquiry about experiences with substitution in the OHS system. Besides the results we have also chosen to present the complete collected material, which we expect can be a catalog of ideas and inspiration for the users to find new ways in their own work with substitution.

In the inquiry we asked for both good and bad experiences with substitution because we believe that the unsuccessful experiences of others can also be of value. We don't intend the suggestions as means for "going out and do the same." The examples are thought as a bank of ideas, where the readers must evaluate if the single idea can be a way to solve the recent problem.

Therefore it is of importance before a substitution to think about such questions as:

- Is the substitution in line with our toxicological knowledge and the technical development?
- Could unwanted connections with other chemicals in the factory happen?
- Would substitution give technical problems with items such as machinery?

As writers of this book we are only carriers of ideas and not responsible considering our readers' use of the ideas in any coincidences.

The systematical use of substitution is a rather new phenomenon and continuous development takes place in the area. The case stories mentioned were collected in 1987 and can as such be considered as a status over the OHS's reach into the substitution area at that time. Already essential new discoveries have been made. A couple of these are mentioned here.

A truly exceptional example of substitution was presented after the publishing of this book. As you can read in the book a lot of work is done to reduce the use of organic solvents. This is primarily caused by brain damaging effects of these materials. The graphical trade is a huge user of this type of chemicals in the cleaning of, for instance, off-set printers. Here the Danish printers discovered that these machines can be washed off with vegetable oils giving the same results. In a lot of printing establishments this means a drastic cut-down in the use of chemicals such as white spirit, kerosene, and toluene. This development has had the positive effect that Danish wash-off producers have reconsidered their product sortiment so they now promote commercial products based on vegetable oils. We hope that great foreign producers of chemical products for the graphical line of business will follow suit.

The Danish code-number system was revised in 1993. Printing inks and fillers were included, among others. The guidelines for determination of code numbers have been revised also. Impurities and monomers have to be accounted for. The executive orders from the Danish Working Environment Services are available in English (Executive Order on Work with Code-Numbered Products (No. 302/1993) and Executive Order on the Determination of Code Numbers (No. 301/1993)).

This book was written for Danish readers in 1987 and the contents have not been altered in this American edition. This means that there are some references to specific Danish conditions and Danish laws that might have been revised. In the text you will find references to literature describing these conditions more closely. Some of the law information is available in English.

We hope that this book can make more people in the United States think about substitution, and we would like to propose to our American colleagues to start their own collection of examples of substitution. We have discovered that the registration itself has further pushed the work with substitution. In Denmark we are anticipating the knowledge about American experiences.

Finally we would like to acknowledge barrister Jon Palle Buhl and HSE (Health and Safety Executive, Great Britain), and Mr. Jonathan M. Russell who have helped to make it possible to publish the book in English.

Have a nice substitution!

Per Filskov
Gitte Goldschmidt
Mogens Kragh Hansen
Lena Höglund
Tea Johansen
Christian Libak Pedersen
Lone Wibroe

TABLE OF CONTENTS

Part I

1. **Definition of the Concept of Substitution** 3

2. **The Working Process of Substitution** 5

   2.1 Formulation of the Problem 7
   2.2 Idea Generation Phase 11
   2.3 Criteria 12
   2.4 Description and Assessment of the Alternatives 12
   2.5 Decision 14
   2.6 Execution 14
   2.7 Control 14

3. **Implements for the Assessment of Substances and Materials** ... 17

   3.1 Threshold Limit Value (TLV) 18
   3.2 Vapor Hazard Ratio (VHR) 21
   3.3 Substitution Factor (SUBFAC) 24
   3.4 MAL Coding (Code Numbers) 26
   3.5 Classification and Labeling 31
   3.6 Toxicity Assessment 36
   3.7 Current Norms 48
   3.8 The Experiences of Other Workers 50
   3.9 Miscellaneous 51

Part II

4. **The Study** 57

   4.1 Background of the Study 57
   4.2 Purpose 58
   4.3 Method 58
   4.4 Material 59
   4.5 Quality of Data 60

4.6 Results	61
4.7 Conclusion	69

5. Examples … 71

5.1 Reading Instructions	71
5.2 Survey	74
5.3 Examples	80
5.4 Index of Professional Groups	168

Bibliography

General Literature on Substitution	171
Handbook Literature to be used in Toxicity Assessments	171
References	173

Part I

1. DEFINITION OF THE CONCEPT OF SUBSTITUTION

Substitution as a method of reducing the effects of hazardous substances and materials has been laid down in order no. 540 of the [Danish] Ministry of Labour concerning substances and materials.[1] This says that "it is not allowed to use a substance or material which may be hazardous to or otherwise reduce safety or health if it can be replaced by a nonhazardous, less hazardous, or less irritant substance or material."

It may be useful to divide substitution into various levels. In this way we may make it clearer what is meant by the word "substitution," and assist in systematizing considerations of substitution in concrete situations.

The common starting point on all levels is that in a given working process a substance or material is used which, by means of substitution, is removed from the working process.

The substitutions may be divided into three levels, according to Sørensen and Styhr Petersen:[2]

	substance	implement	process
Level 1	changed	unchanged	unchanged
Level 2	changed	changed	unchanged
Level 3	changed	changed	changed

Example: Soldering

Level 1: Tin/lead solder → Tin/silver solder

Level 2: Resin flux solder → Water-washable flux solder

Level 3: Soldering → Mechanical joining

From level 1 to 2, the change of the substance has led to a change of the method for cleaning the flux—in this case from immersion in an organic solvent to flushing with water. But the process (removal of flux residue) is still necessary. As mentioned above, the material we began with, on all levels, is removed from the working process.

Considerations of substitution are not only relevant in already existing processes, but are also very useful in the planning of new jobs.

The purpose of a substitution is in all cases to remove or reduce effects from substances and materials which are hazardous to health.

Thus, although many steps serve the purpose of substitution, they cannot be considered as substitutions. This is true when the substance or material is unchanged, but conditions of health are improved by changing the process or the method. This may be the case, e.g., when a manual process of injection is replaced by an automatic process, or by screening or ventilation of the working process.

We have therefore chosen the following definition of substitution:

> Substitution has as its starting point the presence of a hazardous substance or material. Substitution is the removal from the working process of the substance or material which gives rise to considerations of substitution. It may be on one of three levels.

2. THE WORKING PROCESS OF SUBSTITUTION

When substitution is carried out or any other task is performed, it is useful to proceed according to a specific plan or a specific working process, so to speak. The purpose of the working process is to ensure that all aspects of the problem are always taken into consideration, so that the optimal solution may be achieved.

In the following we shall describe a suitable working process for a substitution task. It is obvious that it is not possible in all cases to follow the working process slavishly, but we think that it may serve as a basic plan for most types of tasks. Then it is up to individual resourcefulness to improvise as required. In any case, we hope by this chapter to pass on some practical experiences in substitution tasks.

The working process is developed so that it runs through various phases (see the figure below). Each of the phases deals with a specific aspect of the task, and the order of the phases describes the substitution as a logical process.

There is a tendency to focus on the part of the iceberg in the drawing which is above the water and thus immediately visible—namely, phase 5: the decision. Here something happens. But before the decision, there are a number of phases which, in fact, are just as important as, or more important than, the final decision. Phase 1 especially (the formulation of the problem) is an important phase. Here the field of the substitution is defined. The risk of not proceeding systematically may lead to plunging too quickly into decisions or too quickly regarding solutions as impossible. In these cases, the iceberg is hit and perhaps the ship is not righted before it sinks.

In all phases, it is important to have close contact with the company. Problems ought to be discussed early with each worker that may be involved in substitution. The workers know the working process and its characteristics thoroughly. They may be aware of many things which are almost impossible for the uninitiated to know. Moreover, it is the workers, in the end, who are to use the new product and perhaps change their working procedures, if substitution is to take place. Further, the purpose of the substitution is not least to protect the health of the workers in the end. It is therefore an advantage that they should know the considerations, since it is not always the best psychology to have a final solution handed out.

Supervisors are important partners. They know the purpose of and the premises for the working process, and are familiar with conditions generally in the rest of the company. They may participate in the assessment if the technical requirements *must* be satisfied by the substitution, or if it is possible to modify the technical requirements.

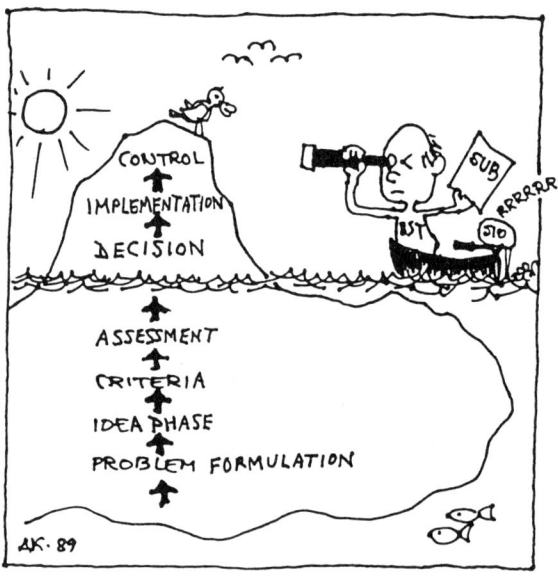

For a detailed study of the working environment, see "Vejledning I arbejdsmiljøundersøgelser" ("Instructions of Working Environment Studies").[3]

2.1 Formulation of the Problem

A good formulation of the problem makes good solutions possible.

> In a laboratory, acetone was used for drying glass equipment. It dried fairly quickly, but it was unpleasant to work with. It was replaced by pneumatic drying—the glass equipment was simply blown dry.

We shall use this very simple example as an illustration of the importance of the formulation of the problem.

A possible formulation of the problem could be: *Acetone is unpleasant to work with—we want to find another substance to replace acetone for drying of glass equipment.* Then ethanol or similar solvents may be taken into consideration.

Another formulation of the problem could be: *We want our laboratory equipment to be dried quickly. Acetone is unpleasant. How else might it be possible to dry the glass equipment quickly?*

A formulation of the problem is an exact description: What is the problem, the background of the problem, and what is to be examined?

The formulation of the problem determines the problems to be dealt with and possible solutions. Therefore the formulation of the problem must take the limitations of the solutions into consideration. Is it, for example, possible to consider an alternative process without the use of a chemical product?

When the problem is formulated it is important not only to consider the product, but the whole working process that the product is part of as well. What are the working functions? What happens before and afterwards? What is the purpose of the working process?

Because the formulation of the problem also must include a good description of working conditions, it is important to visit the company in this phase.

The background for the recognition of the problem may be as diverse as accidents, symptoms, an irritating odor, publicity, or orders from the Danish Working Environment Service.

Thus the elements should be the following:

FORMULATION OF THE PROBLEM

- What is the problem?
- What is the background of the problem?
- What is to be examined?

> - Description of the product
>
>   a) Content of the product
>   b) Toxicology and health effects
>   c) Actual health effects
>
> - Description of the work:
>
>   a) The object or the purpose of the working process
>   b) Description of the execution of the work
>   c) How it is connected with other parts of the work
>
> - Description of exposure

The problem, its background, and the object of the study are dealt with above. In the following we shall go into detail about the remaining three items: descriptions of the product, of the work, and of exposure.

2.1.1 Description of the Product

2.1.1.1 Content of the product

Procure instructions and data from the company, the supplier, etc. and take care that they are *not* obsolete.

Call the supplier/manufacturer and ask for a detailed description of the product. It is often difficult to obtain a complete description of its composition, and it must then be considered whether all details are necessary. Additional information may be obtained from the Danish Working Environment Service's register of products.

> If the content is unknown, it is impossible to assess the product.

2.1.1.2 Toxicology and health effects

Chapter 3 goes into detail about the comparison of substances and materials. One method is to make a so-called "toxicity assessment," but there are a number of other methods. However, toxicity assessment is of particular importance, since it is the basis of all substitution.

What do we know about irritant effects of the product on health?

However, complete information is not always accessible, especially in the case of new products/processes/methods.

> If effects on health are unknown, it is impossible to state the risk of working with the product.

2.1.1.3 Actual effects on health

Make a survey of the actual irritating effects on health.

When and for how long are irritant effects and symptoms experienced—and do they disappear after the day's work, on weekends, or during holidays?

How many people are affected by the irritant effects, and how are the effects distributed in relation to working functions? Is there a connection between working functions in which the products are used and its irritant effects on health?

2.1.2 Description of the Work

2.1.2.1 The object or the purpose of the working process

It is important to make this quite clear in order to ensure a reasonable substitution.

If the purpose of the working process, e.g., is to degrease a material, there are, among others, the following possibilities: organic solvents, bases, or water. Avoid habitual thinking. Is the purpose quite unequivocal? Is degreasing necessary at all?

2.1.2.2 The connection with other parts of the work

Degreasing may be necessary, if the material is to be painted, but maybe only just before the painting, and not as an intermediate phase between each working process.

Substitution of a chemical substance in each phase of the process should be seen in connection with the *total* line of production, with the purpose of decreasing the *total* health hazard.

2.1.2.3 Description of the execution of the work

If the purpose of the working process is, for example, degreasing a material, how does it take place? And when is man involved? How is the working routine characterized? Might the human factor cause accidents in the working process?

Many factors should be taken into consideration in the description.

2.1.3 Description of Exposure

The physical surroundings of the workers are of the utmost importance when exposure to substances hazardous to health is assessed, and it is therefore necessary to know the degree of exposure.

Important factors include:

- For how long each time, how often, and in what quantities is the chemical product used?
- Is there a risk of coming into direct contact with the chemical product—either during the process or by accident?
- Is there a risk of exposure to vapor, smoke, or dust?
- In what state does the pollution leave its source—as vapor, smoke, dust, tiny drops, splashes, or another state?
- What is the temperature of the product during the process?
- What are the flammable and explosive properties of the product?

- What is the size of the surface of evaporation?
- Is there a risk of exposure to accidental compounds of substances and materials (e.g., acid/base aerosols and dust?)
- What are the conditions of ventilation and the exhaust efficiency?
- Are personal safety means used (masks, gloves, etc.)?
- Are breakdown products formed?
- What are the temperature and humidity of the room?
- Are hygiene and sanitary conditions acceptable?
- How is the room cleaned?
- Is there a special risk of exposure by cleaning or repair work?
- What substance(s) and how much leaves the source? This may be measured by a degassing analysis, in which the quantity and the type of gasses of a sample of the product are analyzed.
- How much substance reaches the breathing zone of the worker? This is measured by a working hygienic air pollution measurement.

> If it is unknown how the product is used, it is impossible to assess the risk of working with the product.

The degree of physical work is a further factor which may increase the health hazard. When physical effort is being made, more air is breathed in and there is more exposure to pollution. However, the threshold limit value does not take into account possible hard physical work.

Finally, an assessment of the risk of the actual use of the product may be made, i.e., there may be a total assessment of the conditions of health and exposure.

2.2 Idea Generation Phase

In this phase, information about possible solutions is procured from all kinds of sources. All possible and impossible ideas should be presented. Have the workers any ideas? Has the register of products any ideas? What do they do in other places? Has the supplier any ideas, and what can you imagine yourself? It is important not to limit oneself in this phase to such thoughts as: it is impossible; it is too wild; it is too expensive; it is difficult. On the whole, statements like "It is too..." are forbidden.

In this phase, some examples from this book may be used. One may draw inspiration by leafing through it. It is also important to get to know what others do in similar cases.

Try to find out how the task can be performed without the use of chemicals.

2.3 Criteria

In this phase, the criteria to be used in the assessment of different possibilities of action are chosen. The criteria may be, for example:

- The alternative shall represent a real improvement of health.
- The alternative must not lead to new problems in the working environment, e.g., ergonomic ones.
- The pollution resulting from the substitution may not exceed a fraction of the threshold limit value (e.g., 1/10).
- The alternative must not give rise to immediate irritant effects on the workers.
- The alternative must not make the work more difficult.
- The alternative must not include substances under suspicion of being carcinogenic, teratogenic, allergenic, or causing other chronic damage to health.
- The alternative must not be a depreciation of the external environment.

It is easy to establish more or fewer or more strict criteria.

The criteria should be arranged in order of priority so that one is perfectly aware of the most important ones.

Finally, we have the company's requirements for the alternatives—requirements which must be taken into consideration:

- The alternative must result in 100% degreasing.
- The alternative must not cost more than an additional 50%.
- The alternative must not lead to a decrease of production.

It is important to discuss these requirements. Is it necessary that the alternative be technically as efficient as the "old" product? Is it not sufficient that it is efficient enough? How efficient degreasing must be is, for instance, dependent on the material being painted or welded afterwards. Is there a real foundation for the requirements—or are they a result of habitual thinking: "We have always done it like this."

2.4 Description and Assessment of the Alternatives

Now the possibilities of action produced in the idea generation phase are to be dealt with. A realistic approach is of course needed before time is spent on details. The most wild and quite unrealistic ideas are sorted out. But do not be too pessimistic. Many people have been surprised by what is suddenly possible.

Alternatives are described and assessed in the same way as the original product. See the items in Section 2.1.1, among others.

2.4.1 Assessment of Each Alternative in Light of the Criteria Laid Down

When a total assessment of each alternative of the risk in the concrete working situation has been made, the alternatives are assessed in light of the criteria laid down. Especially in this phase, frequent moving forward and backward between the phases takes place:

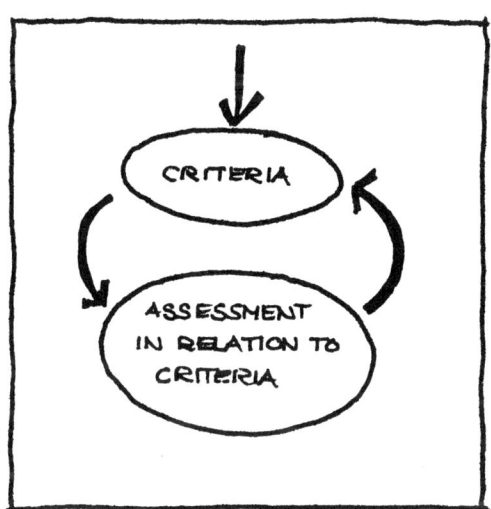

It may be necessary to reestimate the criteria and allow new parameters for the assessment of the products.

When estimating whether the product satisfies the technical requirements, one is dependent on the promises of the supplier or of the technicians of the company. To avoid that, the Occupational Health Service is made responsible for the technical properties of the products. It is possible to let the suppliers be in direct dialogue with the company, so that the Occupational Health Service does not promise miracles.

2.4.2. Assessment of the Alternatives Mutually and in the Light of the Criteria Laid Down

Against the background of the above-mentioned work, it should not be too difficult to choose among the possible alternative solutions, so that one or more possible alternatives may be proposed.

2.5 Decision

By this phase we mean a decision together with the company.

When the decision has been made, it may be tempting to finish the work, but we do not think that a good substitution is brought to an end before it has been tested in practice and assessed.

After much consideration, one or a couple of alternatives are often left. In consultation with the company, the best alternative may then be found. This part of the process may be prolonged and complicated, since all parties (e.g., management, workers, safety board, and superiors in the department) perhaps do not agree on the basis of the assessment. Are unmentioned criteria having an influence? It may be necessary to mobilize all one's gifts of persuasion.

2.6 Execution

At this time it is very important again to include the workers who are to carry out the work the substitution concerns. Discuss the choices with them. The desire for substitution does not always spring from those concerned. Therefore it may be a good idea to discuss advantages and disadvantages of the new product and why it is useful to substitute.

Define a period when the product is to be tested and talk with the worker(s) about the things to be noticed in this test period.

2.7 Control

After a test period, the following may serve as examples of assessments:

- Does the product satisfy the criteria that were initially laid down?
- Have effects on health diminished?
- Are there signs of new effects on health from use of the product?

THE WORKING PROCESS OF SUBSTITUTION

- Have new problems of the working environment arisen, e.g., ergonomic ones?
- Have changes of personal exposure occurred after the introduction of the new method? If there is any uncertainty, recourse may be taken to control measurements.
- Are the company and end user satisfied with the new product with respect to production?
- Have all problems been solved, or should the proposed solution perhaps be combined with, e.g., exhaust ventilation? Should other products be tried—or should a new start be made?

3. IMPLEMENTS FOR THE ASSESSMENT OF SUBSTANCES AND MATERIALS

When a substitution of a substance or a material is to be made, it is important to make a qualified assessment of the health properties of the substance or the material, including the relevant alternatives. It is thus ensured that the substitution actually becomes an improvement to health conditions.

There are a number of tools for such an assessment. Some of them will be dealt with in the following:

1. Threshold limit value (TLV)
2. Vapor hazard ratio (VHR)
3. Substitution Factor (SUBFAC)
4. MAL coding
5. Classification and labeling
6. Toxicity assessment
7. Current norms
8. The experiences of other workers
9. Miscellaneous

As mentioned in Chapter 2, the toxicity assessment has a special position: it is impossible to talk about the risk of working with a product if health hazards are unknown and some kind of toxicity assessment has not been made for the purpose of substitution.

It is also important to note that substitution implies a total assessment of the working situation, the chemical substances, etc. It is seldom possible to substitute by sitting at a desk applying a single implement; several of the implements often have to be used at the same time—and it is always necessary to make a kind of toxicity assessment, so that the health hazards of using the substances and materials have been studied. On the contrary, one tool may very well be used as the decisive cause of substitution.

The following example illustrates the utility and the strong and weak sides of the above-mentioned tools.

Example: Substitution for PVC adhesives

A company addresses the Occupational Health Service because some workers suffer from headache and fatigue when bonding PVC materials. The company wants the Occupational Health Service to procure an adhesive which is better for the environment.

The adhesive product used contains butanone (methyl ethyl ketone, MEK), tetrahydrofuran (THF), and dimethylformamide (DMF). Gloves and an exhaust box are used in the workplace.

A study of alternative products favors an adhesive based on *N*-methyl-pyrrolidone (NMP). The product is chosen, among other things, on the basis of measurements of evaporation from different adhesives. These measurements were made for one of the manufacturers by an independent laboratory.

Soon after the introduction of the new adhesive, a worker's arms begin to tingle, which results in an illness report. At the same time, quality problems concerning the bonding arise, and the setting time is increased.

The company rejects the substitution and, instead, the materials are changed, so that they are assembled mechanically with a technical result which is just as satisfactory.

3.1 Threshold Limit Value (TLV)

3.1.1 The Tool

Threshold limit values are administrative norms of air pollution laid down by the ACGIH and the Danish Working Environment Service. TLVs of approximately 600 substances are found in the instructions of the Danish Working Environment Service about TLVs of substances and materials.[4] According to these instructions, TLV is an expression of the concentration a

ASSESSMENT OF SUBSTANCES AND MATERIALS

substance is not allowed to exceed, calculated as a time-weighted average during an 8-hour working day.

TLV is determined on the basis of the accessible documentation of health, and this is considered from a technical and economic point of view. These considerations are made by the Working Environment Council, and the TLVs are thus an expression of agreements between management and labor.

TLV may be determined in consideration of very different effects. Approximately 40% of the TLVs are thus determined on the basis of the irritant effects of the substances in question.[5]

The TLV list of substances and materials which are hazardous to health is far from complete. Therefore, a guideline list of organic solvents with the so-called "tentative threshold limit values" has been made. These values have not been dealt with by the Working Environment Council, and the documentation is generally not as satisfactory as for the substances in the TLV list itself.

In the TLV list, notice is given of the changes and additions which the Danish Working Environment Service considers as documented. Changes and additions are entered on the subsequent revised list to the extent management and labor agree.

The TLV list was planned to be revised every second year. However, there has been a considerably longer span of time between the latest revisions.

3.1.2 Substitution by Means of TLV

As is apparent from the above, the TLV not only expresses the health hazard of a substance, but the value may be used with care as a guideline when two substances are compared. As a rule, the lower the TLV, the greater the risk of using the substance concerned.

3.1.3 Example

The PVC adhesives of the introductory example are assessed in the following way:

	Product 1			Product 2
	THF	MEK	DMF	NMP
Threshold limit value (ppm)	200	100 H	10 H	100
Quantity (vol%)	55	17	7	70

Only vol% of the volatile substances are indicated. The rest of the product consists of dry matter (PVC powder).

The comparison of the two products by means of TLV is not easy. There are no formulas indicating the "resulting" TLV of a product which consists of several single substances, each with its own TLV (as, e.g., product 1). Approximately half of product 1 consists of a substance with a TLV that is higher than that of product 2. However, there is not as much DMF in product 1, which has a considerably lower TLV than NMP. On the basis of TLV alone, it is not obvious which product to choose.

3.1.4 Discussion

There are today 10,000 to 50,000 chemical substances and materials estimated to be in the Danish market.[27] As mentioned above, TLVs have only been determined for approximately 600 of the substances. When two products are to be assessed, it is therefore often the case that they contain substances without TLVs. In this case a direct comparison is of course impossible. TLV is an easy and quick method only if the substitution is sufficiently simple, which means in practice that the alternative products only contain one hazardous substance each and that the TLV has been determined for both of them.

However, the method has some serious limitations:

> The quality of the toxicological documentation is variable. Some substances have been thoroughly studied—whereas little information is available on the effects of others.

At the same time there is a certain delay before the toxicological knowledge about the substances is observed in the determination of TLV.

It is not uncommon that organic solvents, which are usually considered damaging to the central nervous system, have TLVs determined on the basis of irritant properties.

When the TLV is changed, the whole basis of the substitution is changed. In some cases rather dramatic changes of TLV take place. Reductions to 1/10 of the former value are not unknown. The 1985 list of notification shows that the TLV of 1,3-butadiene is considered for reduction to 1/100 of its former value. If the assessment of the substitution is based on a comparison of TLVs, the basis may be changed drastically when new TLVs are determined.

Furthermore, substitution by means of TLV is based on an assumption that TLVs are comparable. This is not always the case. How is, for instance, a TLV-based knowledge of long-term effects on the central nervous system compared to a TLV-based knowledge of irritation of the mucous membranes?

ASSESSMENT OF SUBSTANCES AND MATERIALS

In the example of the PVC adhesives we have also seen that it is impossible to compare directly the TLVs of two products when these contain several substances with separate TLVs.

It is important to note that TLV only informs us of air concentrations, and only to a very limited extent deals with other types of exposure. The skin notation—in which substances that can be absorbed by the skin are marked by H—is an attempt to deal with this. However, *by no means all* substances that can be absorbed by the skin are marked by the notation, e.g., NMP is known to be able to penetrate the skin.[6]

Finally, there may be considerable differences of the physical and chemical properties of the substances, e.g., the rate of evaporation is central, since it determines the quantity of the substance one is exposed to when breathing.

This means that a substitution in which a substance with a double TLV is chosen may result in increased health hazard, if the substance, for example, evaporates 10 times faster.

This has been remedied by the tools "vapor hazard ratio" and SUBFAC, which are described in the following sections.

3.2 Vapor Hazard Ratio (VHR)

When the TLVs are solely a "toxicological index," VHR expresses the probability of exceeding the TLVs. Thus it has been taken into consideration that chemical substances behave in different ways in the working situation, which means here that they are more or less volatile.

3.2.1 The Tool

For pure substances, VHR is defined:

$$VHR_i = C_i/TLV_i$$

C_i and TLV_i are, respectively, the "pure component equilibrium concentration," and the threshold limit value of substance I. C_i indicates the maximum concentration of the substance that occurs in air; it is extremely dependent on temperature (higher temperature of air and product ≥ quicker evaporation and increased equilibrium concentration). VHR_i expresses how many times, at maximum, the TLV can be exceeded.

VHRs of different substances may be compared: the lower the VHR, the fewer times the threshold limit value can be exceeded. This is often expressed as: the lower the VHR, the lower the probability of exceeding the threshold limit value.

C_i is often indicated as units of pressure (e.g., mmHg, Pa, atm) and is then called the "pressure of saturated vapors," vapor pressure, and the like.

If C_i is indicated as mmHg, the conversion to ppm takes place by means of the formula:

$$C_i(\text{ppm}) = P_i(\text{mmHg}) \cdot \frac{1000000}{760}$$

Even if VHR is only defined for pure substances, it is often applied to compounds:

$$\text{VHR} = \frac{V_1 \cdot C_1}{\text{TLV}_1} + \frac{V_2 \cdot C_2}{\text{TLV}_2} + \ldots + \frac{V_x \cdot C_x}{\text{TLV}_x}$$

The compound here consists of x substances, each with a TLV and a saturated vapor pressure (the equilibrium concentration). Each substance is found in the compound in the quantity V (the weight percentage).

Such an increase of VHR is used for the calculation of MAL factors. (See MAL coding in Section 3.4.)

3.2.2 Substitution by Means of VHR

Whether the TLVs are exceeded is, of course, dependent on the handling of the given product.

Because VHR expresses the probability of the TLVs being exceeded, a product with a low VHR will be preferred to one with a high VHR, after all.

3.2.3 Example

The substitution for PVC adhesives is performed here by means of VHR calculations.

	Product 1			Product 2
	THF	MEK	DMF	NMP
Threshold limit value (ppm)	200	100	10	100
Equilibrium conc. (ppm)	170,000(8)	127,000(7)	3,751(8)	525(9)
Quantity (vol%)	55	17	7	70

Only vol% of the volatile substances are indicated. The rest of the product consists of dry matter (PVC powder).

ASSESSMENT OF SUBSTANCES AND MATERIALS

VHR calculations show the following result:

VHR (product 1) = 710   VHR(1)/VHR(2) = 178
VHR (product 2) =   4

The ratio between the VHRs expresses the ratio between the number of times the threshold limit values can be theoretically exceeded.

The VHR calculation is to the advantage of product 2: its maximum excess over the threshold limit value is here 178 times below that of product 1.

Please note also that C_i does not indicate the rate of evaporation of a substitution, but merely the largest concentration of the substance in air. However, C_i and the rate of evaporation are roughly proportionate. If Ç is doubly increased, the rate of evaporation will be increased approximately three times (10).

As it appears from the example, 7 indicates the equilibrium data for only 1 of the 4 substances, and it has been necessary to refer to other data sources.

3.2.4 Discussion

Contrary to the TLV, the VHR method takes into account different evaporation rates of chemical substances.

VHR uses the threshold limit values as an expression of the effects hazardous to health. The weaknesses of TLV are therefore transmitted to VHR. The method may lead to incorrect results, if products with very different components are concerned.

The difference may concern the polarity of the components, which among other things expresses the mutual, miscible properties. (Benzene is a nonpolar solvent, whereas water and ethanol are polar solvents.)

The results of VHR calculations of an "aqueous" product containing residues of organic solvents, e.g., toluene and benzene, will be far from actual conditions.

VHR may seldom be used as the only implement in a substitution, in that it does not take into consideration whether the chemical substances can be absorbed by the skin or breathed in as aerosols, which damages are entailed, etc.

An obvious question is: by how many times VHR must be lower in order to make the substitution successful? Is it "enough" to replace with a product with half the VHR? Or must the proportion be above 100? It must be remembered that VHR is a theoretical number which does not take into consideration how the products are used in the workroom. It is therefore reasonable to require a considerable difference of VHR to be sure that the substitution works well.

In Section 4.6, the VHR is used as a clear index of the examples referred to.

3.3 Substitution Factor (SUBFAC)

As mentioned above, the VHR may lead to erroneous conclusions if the product is made of components with different solvent properties.

SUBFAC is a tool that gives more realistic results for such products. The method goes further into detail about the mutual behavior of single substances of the product. Calculation is therefore also more complicated than in the case of VHR, and it can only be practiced by means of a computer.

The theoretical background of the method is briefly described here (in detail in [11]).

3.3.1 The Tool

The SUBFAC of a substance is defined as the "strength of the source" divided by the threshold limit value of the substance:

$$SUBFAC = phi_i/TLV_i$$

The strength of the source, phi, is given in $mol/s/m^2$, i.e., the quantity of the substance which evaporates per second per square meter of the surface of the liquid. The definition of SUBFAC shows its similarity to VHR.

A couple of presuppositions reduce the problem to the calculation of:

$$SUBFAC_i = gamma_i \cdot x_i \cdot C_i/TLV_i$$

where C_i is the equilibrium concentration, which is also known from the VHR, x_i is the mol fraction (another way of describing the quantity of the liquid made up by substance I), TLV_i is the threshold limit value of substance I, and $gamma_I$ is the so-called coefficient of activity.

These presuppositions result in SUBFAC no longer describing the rate of evaporation, but solely the "equilibrium concentration," as accurately as VHR.

The new entity separating SUBFAC from VHR is the gamma activity coefficient. This entity has to be calculated by computer.

If a product contains several components (they often do so), the total SUBFAC of the product will be:

ASSESSMENT OF SUBSTANCES AND MATERIALS

$$\text{SUBFAC} = \sum_i \frac{\text{gamma}_i \cdot x_i \cdot C_{i_i}}{GV_i}$$

The procedure of summing up from single components of the product is recognized from VHR. Note that if all coefficients of activity are equivalent to 1, SUBFAC is identical to VHR.

3.3.2 Substitution by Means of SUBFAC

Two products are substituted by choosing the product with the lowest SUBFAC value.

However, just as is the case with VHR numbers, it is difficult to interpret each SUBFAC number: it is not certain that the threshold limit value is exceeded, even if SUBFAC shows this (dependent on working situation, design of the machine, ventilation, etc.).

Nevertheless, generally, the smaller the risk of exceeding the threshold limit values, the lower the SUBFAC of the product.

3.3.3 Example

The substitution for PVC adhesives results in the following SUBFAC figures:

SUBFAC (product 1) = 921
SUBFAC (product 2) = 5
SUBFAC(1)/SUBFAC(2) = 184

The calculations have been made by Lars Justesen, of the National Institute of Occupational Health, Denmark.

As in the VHR example, product 2 is still to be preferred.

3.3.4 Discussion

SUBFAC leads to surer results than VHR when products are dealt with which contain several chemical substances. When VHR presupposes all gamma = 1, it has been found by SUBFAC that gamma may vary between 0.4 and 10,000,000!

SUBFAC may lead to incorrect results if the product contains components that are deliberately intended to reduce the evaporation.

If the recipe of the product is changed, it is not immediately possible to calculate a new SUBFAC on the basis of the old one, because the coefficient of activity is dependent on the quantities of each substance in the product.

As mentioned above, the coefficients of activity are dependent on the composition of the product. SUBFAC will therefore be changed if the product is changed during the evaporation—as is the case of, for example, painting, by laying the product (the paint) out in a thin film.

TLV is used for assessments of effects hazardous to health. All the limitations of TLV are therefore transmitted to SUBFAC. Moreover, SUBFAC cannot be used if the products can be absorbed through the skin, or if they are aerosols (liquid vapors) in the air.

As is the case with VHR, the alternative product must have a considerably lower SUBFAC in order to make a proper substitution. There is thus no reason to substitute a product having only half the SUBFAC.

It is time-consuming to calculate SUBFAC, because the computer calculations are not within the scope of the Occupational Health Service. However, there are plans to develop a simpler PC computer version of the calculating program, so that the calculations can be made by each Occupational Health Center. Furthermore, it is planned to calculate SUBFAC of all relevant products and include them in the database of the Register of Substances and Materials.

3.4 MAL Coding (Code Numbers)

If an expression is added to VHR of the degree of hazard of the product to skin and eyes by inhalation of aerosols and dust of the product, and if the product is swallowed, we obtain the MAL code system. The code system is intended for paints, but is also used in practice for a number of other types of products, e.g., adhesives and cleaning liquids.

The code number of each product is composed of two figures: **X-Y**. The figure before the hyphen refers to the health hazard by inhalation of vapors of the product, and may have the values 00, 0, 1, 2, 3, 4, and 5.

The figure after the hyphen refers to health hazard by contact with skin and eyes and by inhalation of spraying mist, dust, etc., and may have the values 0, 1, 2, 3, 4, 5, and 6.

3.4.1 The Tool

In this section, the calculation of the code number of a product is dealt with briefly. Reference is made to the order of the Danish Working Environment Service concerning coding of painting products.[12] See, furthermore, Reference 13.

3.4.1.1 Calculation of the Figure before the Hyphen

First the "MAL factor" is calculated for each component in the product. (MAL = amorent of air needed for occupational hygiene).

$$\text{MAL - factor} = \frac{k \cdot 10000}{\text{TLV}}$$

The MAL factor shows how many cubic meters of air are necessary to dilute the polluted air when 10 g of the substance in question in a pure state is used to avoid exceeding the threshold limit values. "k" may assume a number of values: the higher the rate of evaporation (or equilibrium concentration), the larger k is. The MAL factor is thus a calculating factor taking into consideration the threshold limit value and the volatility of the substance.

In Supplement 1 to the Reference 12, the MAL factors are calculated for a number of substances to save effort. On the basis of the MAL factors, the MAL of the product is calculated as a sum of the single MAL factors (MALf):

$$\text{MAL} = D \cdot (\text{MALf}_1 \cdot V_1 + \text{MALf}_2 \cdot V_2 = \cdots + \text{MALf}_x \cdot V_x)$$

D is the density of the product (kg/L) and V_x is the percentage by weight of component x in the product.

Thus MAL, may be explained as: *Number of m^3 fresh air necessary to dilute the air polluted by 1 liter of the product so that the threshold limit values are not exceeded.*

On the basis of MAL, the figure before the hyphen may be found by means of the following table:

MAL	The figure before the hyphen
0 m³/l ≤ MAL ≤ 30 m³/l	00-
30 m³/l < MAL ≤ 100 m³/l	0-
100 m³/l < MAL ≤ 400 m³/l	1-
400 m³/l < MAL ≤ 800 m³/l	2-
800 m³/l < MAL ≤ 1600 m³/l	3-
1600 m³/l < MAL ≤ 3200 m³/l	4-
3200 m³/l ≤ MAL	5-

3.4.1.2 Calculation of the Figure After the Hyphen

In the above-mentioned supplement the figure after the hyphen is indicated for a number of substances. In addition, a percentage by weight limit is indicated. If the product contains several substances with the same figure after the hyphen—but in quantities below their respective percentage by weight limits—the product is given the number concerned after the hyphen, if

$$P_1/G_1 + P_2/G_2 + ... + P_x/G_x > 1$$

where P_x and G_x are, respectively, percentage by weight and the percentage by weight limit of substance x. If the product contains a substance with a higher number after the hyphen and in a quantity above the threshold limit weight-percent of the substance, the product is, of course, given the number of this substance after the hyphen.

The various figures after the hyphen refer to (see the exact text in Reference 12):

0   Water
1   Substances/products which are not hazardous to health, except by inhalation of spraying mist, dust, etc.
2   Substances/products which are not hazardous to health, except if swallowed and by inhalation of spraying mist and dust
3   Damages to skin and eyes, and by inhalation of spraying mist
4   Corrosive substances/products
5   Substances/products which may cause allergic reaction
6   Toxic substances/products

3.4.2 Substitution by Means of MAL Coding

The code number of a paint may serve as an easy basis for substitution, in that higher numbers indicate an increased health hazard. On the whole, a low code number is preferable: 2-1 is better than 5-3. This example is obvious; it is more difficult when the assessment is to be made for or against 3-5 in relation to 5-3. However, the principal rule is that the figure before the hyphen is decisive (3-5 is chosen instead of 5-3); but see the comments in, e.g., Reference 13.

In the concrete substitution it is worth noting how the product is used; one should not merely substitute blindly on the basis of MAL codes.

3.4.3 Example

The substitution for the PVC adhesives is carried out here by means of MAL coding.

ASSESSMENT OF SUBSTANCES AND MATERIALS

	Product 1			Product 2
	THF	MEK	DMF	NMP
MAL factor	24	32	230	8
Figure after hyphen	-1	-1	-6, -3	-1
Threshold limit wt%	0	0	≥ 5, 0,5–5	0
Quantity (vol%)	55	17	7	70

The density here is fixed at 1 kg/L.
MAL of the two products is:

$$MAL(1) = 24 \times 55 + 32 \times 17 + 230 \times 7 = 3474$$
$$MAL(2) = 8 \times 70 = 560$$

The figure before the hyphen is therefore:

Product 1: **5-**; product 2: **2-**

The figure after the hyphen is -6 for product 1, because the limit percentage by weight of DMF is above 5.

For product 2, the figure after the hyphen is -1.

MAL codes: Product 1: **5-6**; product 2: **2-1**

The MAL codes show that it is obvious that product 2 should be used instead of product 1.

3.4.4 Discussion

One of the intentions behind the method is that it may be used alone. Or, as formulated in the ministerial order of 1982 concerning professional painting:[14]

> § 6. The use of a product is not allowed if it can be replaced by another product which, according to the code number, is less hazardous or less irritant.

Furthermore, the method is easy to use but it has some limitations:
The figure before the hyphen is based on the VHR and may therefore cause trouble with compounds of widely different types of substances. (See the comments under VHR.)

Subsupplement 1 to the ministerial order on coding of paint codes is based on 1981 values of the threshold limit values. Even if the threshold limit values have since been lowered, the MAL factors indicated in this subsupplement must be used. For convenience and orientation the "correct" figure before the hyphen may then be calculated—i.e., the figure which, instead of building on 14-year-old threshold limit values, is based on the threshold limit values of today.

The MAL codes take the actual work situation into consideration only to a small extent: is the paint product, for example, used in a room for 8 hours, or are 10 ceilings painted quickly one after another? Calculations prove that the MAL codes may show different results for substitution when a comparison is made with the air concentrations of solvents painters may be exposed to when they inhale in the two working situations.[15]

All substances in subsupplement 1 except formaldehyde, ammonia, and acetic acid shall be included in the MAL factor, no matter how small a quantity is in the product (i.e., content > 0%). For the two first substances, the limit is, respectively, 0.1% and 0.2%.

The figure before the hyphen is thus dependent on the concentration of the substance being below or above these limits. The limit is initially subtracted from the concentration of the substance in the product. The MAL code is calculated on the basis of this difference. The reason is that substances are bound "chemically" in the product.

Example: **0.10% formaldehyde in water is coded 00-1** (calculated as if there were no formaldehyde in the product).
0.12% formaldehyde in water is coded 1-3 (calculated as if there were 0.02% formaldehyde in the product).

There is, of course, no objective chemical reason for this gap in the code number: only a slight improvement of the working environment is achieved by replacing a 0.12% solution of formaldehyde with a 0.10% solution. Further, some manufacturers are not very good at calculating the MAL codes correctly.

A new ministerial order is being drafted concerning coding of paint products. Some of the growing pains the former order suffered should be cleared away. Among other things, it is expected that attempts will have been made to solve the above-mentioned problems of formaldehyde and ammonia, which should result in many paint products containing the two substances being coded more strictly than they are today. Moreover, the list of MAL factors of the ministerial order is expected to be updated in accordance with existing threshold limit values.

ASSESSMENT OF SUBSTANCES AND MATERIALS

It has not been possible to press the Directorate of the Danish Working Environment Service into revealing specific changes of the new order.

3.5 Classification and Labeling

The rules of classification and labeling of chemical substances are described in the Danish Environmental Protection Agency's order no. 662 from 1987[16] and the accessory list: "List of Hazardous Substances".[19]

The classification and labeling of a chemical substance or product consist of, among other things:

- Hazard class (with warning symbol)
- Risk and safety statements (R- and S-)
- Indication of the names of the hazardous substances in the product that are the cause of the classification.

The labeling shall be indicated on a label on the packaging and in the supplier's instructions.

The rules may be difficult to understand, but then, it must be remembered that the user himself is not to label and classify during the substitution process. This is the manufacturer's responsibility, and it is also his responsibility if a substance which is not in the list of hazardous substances is introduced to the market: then the manufacturer shall document the labeling and classification by means of a number of references in the literature (e.g., experiments on animals). If the substance is new in the Danish market and the documentation is not sufficient, the manufacturer himself shall see to it that the required experiments on animals are carried out.

Furthermore, two orders are worth noting: the order of labeling, etc., of solvents[17] and the order of labeling, etc., of paint articles.[18] In the following description of labeling and classification, reference is solely made to the "principal order"; the two latter special orders are dealt with in Section 3.5.4.

3.5.1 The Implement

3.5.1.1 Division into hazard classes

The labeling system comprises ten hazard classes with corresponding symbols:

Explosive E | Extremely Flammable Fx | Flammable F | Oxidizing O

Very Toxic Tx | Toxic T | Harmful Xn | Corrosive C | Irritant Xi

The hazard classes are briefly called E, Fx, F, O, Tx, Xn, C, and Xi. The tenth hazard class, called "Combustible," has no symbol.

It appears that five of the classes refer to health hazards, the rest to chemical-physical properties. Thus a chemical substance may be placed in two hazard classes (e.g., Xn, O).

In the "List of Hazardous Substances" a number of pure chemical substances have already been classified and assigned R- and S-statements. In the following, the criteria of placement in the hazard classes are dealt with. These criteria are also used when new substances which are not entered on the above-mentioned list are estimated.

Very toxic (Tx)/Toxic (T)/(Xn)

Placement of a chemical substance in these hazard classes is based on the toxicity of the substance—especially the acute toxicity effects—but also by considering long-term effects (e.g., allergy and carcinogenic effect). The acute toxicity is based on the so-called LD_{50} and LC_{50} values:

Hazard class	Rat, through the mouth, mg/kg	Rat/rabbit, through the skin, mg/kg	Rat, inhalation mg/m^3/4 h
Tx	$LD_{50} \leq 25$	$LD_{50} \leq 50$	$LD_{50} \leq 500$
T	$25 < LD_{50} \leq 200$	$50 < LD_{50} \leq 400$	$500 < LD_{50} \leq 2000$
Xn	$200 < LD_{50} \leq 2500$	$400 < LD_{50} \leq 2000$	$2000 < LD_{50} \leq 20000$

Concerning LD_{50} and LC_{50}, see Section 3.6.1. The lower the LD_{50} or LC_{50}, the higher the degree of toxicity, and thus the stricter the hazard class.

ASSESSMENT OF SUBSTANCES AND MATERIALS

Chronic damages to health must also be dealt with in studies of labeling or of damages which appear only under persistent influence. Supplement 1 to the order, items 51 to 53,[16] e.g., describe how substances should be classified, if they are carcinogenic, mutagenic (may damage the genes), or teratogenic (may cause birth defects).

Please note that one is not allowed to classify a product more restrictively than provided by the order.

Corrosive (C)/Irritant (X_i)

These two hazard classes are used when a substance may damage skin and/or eyes in different degrees and the experimental basis of the hazard classes is standardized in the same way as the LD_{50}/LC_{50} regulations. (See Section 3.6 on toxicity assessments.) A substance shall be classified as corrosive if the standard method results in "deep damages to the skin tissue, on a single animal." A substance shall be classified as locally irritant if the standard method results in "irritation/edema/crusts" on the skin, or damages to the eyes (irritation, edema, etc.).

*Extremely flammable (Fx)/Flammable (F)/Combustible/
Oxidizing (O)/Explosive (E)*

In these hazard classes, the chemical-physical properties of the substances determine the place assigned to a substance—whether the substance is flammable, and explosive as well, etc.

A substance is classified as Fx, F, and combustible, if its flash point is below 0°C, between 0 and 21°C, and between 21 and 55°C, respectively. A couple of other criteria belong to the hazard class F.

A substance is placed in the hazard class "explosive," if it explodes in contact with fire or if it is sensitive to bumps and frictional heat (e.g., trinitrotoluene (TNT) and several other trinitro compounds).

A substance is placed in hazard class O if it sustains fire (e.g., all peroxides, liquid oxygen, concentrated nitric acid).

The flash point is defined as the lowest temperature at which a mixture of air and the evaporated substance may spontaneously catch fire—or in other words: the more volatile a substance is, the lower the temperature necessary to obtain a given quantity of flammable vapor in the air, and the lower the flash point.

3.5.1.2 Risk and safety statements

Supplement 1 of the order[16] provides information about the risk and safety statements a given hazard class implies. As mentioned above, these must be indicated together with the warning symbol on a label on the packaging. We shall not deal further with these rules here.

3.5.1.3 Labeling of the product

If all the chemical substances of a product are entered on the list of hazardous substances, all the above-mentioned considerations of hazard classes may be spared, since there are in this case rules of calculation for the determination of the hazard classes. Reference is made to Subsupplement 1 to order 662;[16] e.g., a product Xn is classified according to the following calculation, if there are Tx, T, or Kn substances in the product:

$$V\text{-}\%(\text{substance 1}) \times I_1 + V\text{-}\%(\text{substance 2}) \times I_2 + ... + V\text{-}\%(\text{substance x}) \times I_x > 100$$

V-% is the weight percentage of the substance in the product. It is a so-called index figure, which may have the values 500, 100, and 10, dependent on substance x being classified as Tx, T, or Xn.

It follows that a product with an Xn substance dissolved in water is classified as Xn if the concentration is above 10%, and it is not classified if it is below 10%.

3.5.2 Substitution by Means of Classification and Labeling

In substitutions, classification and labeling cannot be used as systematically as, e.g., MAL coding, VHR, or SUBFAC. Preferably, products should be used that are not classified at all, since these should not be hazardous. See the discussion in Section 3.5.4.

Nevertheless, there are some general "rules":

- Xn is preferred to T and Tx.
- Xi is preferred to C.
- Combustible is preferred to F and Fx.

3.5.3 Example

The following table shows the classification of the components of content in PVC adhesive products.

ASSESSMENT OF SUBSTANCES AND MATERIALS

	Product 1			Product 2
	THF	MEK	DMF	NMP
Vol%	55	17	7	70
Hazard class	F, Xi	Xn, F	Xn	Xi
R-statements	11-19-36/37	11	20/21-36	36/38
S-statements	16-29-33	9-16-23-33-48	26-28-36	41

Please note: Our calculations are based on volume-% and not weight-%.

Classification: Product 1: Xn and F; product 2: Xi.

Owing to the hazard classes, product 2 is preferable, but it is also preferable because of the marking of the single substances.

3.5.4 Discussion

Classification and labeling of chemical substances and products in principle give information about all damaging effects. They may either be read roughly from the hazard class or in more detail by means of the R-statements. However, a number of circumstances make the method rather useless as a substitution tool.

A product which is not to be marked is not necessarily harmless, e.g., several "common" cleaning agents do not have to be marked. (A product shall contain more than 10% of an Xn substance—e.g., xylene and toluene—before it has to be marked Xn.)

The labeling does not always indicate the actual damaging effects of a chemical substance. Several organic solvents are only classified according to their flammable properties and irritant effects (e.g., in the case of acetone, isophorone, and other ketones), and not according to their long-term effects on the brain and the nervous system. But in the "List of Hazardous Chemicals,"[19] risk statement no. 48 is added to 13 organic solvents: "Serious health hazard after long-term influence." This is an allusion to permanent damages to the brain and nervous system.

Approximately 30 substances have the R-statement "carcinogenic." The list of the Danish Working Environment Service of substances considered to be carcinogenic (indicated in the instruction about threshold limit values, Reference 4) includes more than 200 substances with this predicate.

Products may not be correctly labeled. It is up to the manufacturer/importer to judge if the product shall be labeled and classified, and if so, how. There is not presently the capacity to control all substances and products, and the controlling authority does not have enough resources to check whether all products are correctly labeled and classified. If one comes across a product with incorrect labeling, it should be reported to the Danish Environmental Protection Agency.

A number of products are differently classified and labeled if they are used as solvents or paint materials,[17–18] e.g., the limit of labeling a diluted Xn substance as Xn is 10%, unless it is used as a solvent. Then the limit is between 5 and 50%, depending on the substance. The limit for methylene chloride is 50%. A solvent may therefore contain five times as much methylene chloride as another product before it is classified as Xn.

In Reference 20—order no. 181 about materials containing volatile substances, including organic solvents—it is required that these substances be indicated by names in the working hygienic instructions if they are in quantities above 1%. But this does not result in changes of the classification and labeling.

3.6 Toxicity Assessment

Whereas VHR, SUBFAC, and MAL coding were based on the threshold limit values as an index of effects hazardous to health, the toxicity assessment is based directly upon the assessment of data which describe the health hazards.

Toxicology means the "science of toxic substances," and toxicological data may be based on widely different studies and experiments. The studies may be laboratory experiments on animals and microorganisms, or they may be epidemiologic. However, the purpose of the toxicity assessment is the same, namely, to collect these data and this information for a product assessment of the effects which are hazardous to health.

As mentioned above, data vary, but sometimes luck may lead to comparable data of substances and products; this is the exception rather than the rule.

The toxicity assessment thus describes the damages to health which a product may inflict. Thus, the assessment has a special position among the tools in this chapter, because it is necessary to make some toxicity assessment or other to be able to substitute at all. In brief, to be able to substitute, one has to know the effects that are hazardous to health.

A description follows of the data to be looked for in a toxicity assessment, and of the experiments behind the data.

3.6.1 The Tool

A toxicity assessment may be divided into two phases:

- Collection and processing of toxicological data of the components of the product. This phase results in an assessment of the toxicity of the product.
- Determination of the risk of the health hazards from using the product.

The following classification of the toxicological data may be useful for clarification:

1. Absorption and excretion, conversion (metabolism), accumulation
2. Acute toxicity
3. Skin and eye effects
4. Mutagenic effects
5. Cancer
6. Damage affecting reproduction (damage to eggs and sperm cells, teratogenic effects, damages of direct transmission of substances by breast milk, etc.)
7. Allergy
8. Other chronic damage to health

3.6.1.1 Absorption and excretion, conversion (metabolism), accumulation

Chemical substances may be absorbed by the body in three different ways:

- Through the lungs (by inhalation)
- Through the gastrointestinal tract
- Through the skin and the mucous membranes

Even if the chemical substances used on the job are not directly eaten or drunk, a number of substances are often absorbed by the gastrointestinal tract. This is the case for dust and aerosols, for example, which—when they are stopped in the mucous membranes of the respiratory passages—are transported upward to the gullet to be swallowed. Studies have proved that lead, which is used in industry, is often absorbed through the gastrointestinal tract. The reasons are not only the above-mentioned, but also that lead sticks to the fingers and the clothes when lunch is consumed and is thus literally eaten.

The structure of the chemical substances is decisive in the absorption in the body; the size of dust particles and aerosols decides whether they reach the lungs or stop in the upper respiratory passages.

Finally, the chemical structure of the substances determines the way the body converts them before they are excreted, which may take place more or less slowly. Certain substances may even be accumulated (stored) in the body, e.g., lead, fluorine, and strontium, all of which are accumulated in the osseous tissue.

Some organic solvents are especially easily accumulated in the fatty tissue (e.g., trichloroethane). There is much fatty tissue in the brain and the nerves.

Data on the absorption and excretion of chemical substances are often not comparable.

The half-life shows how much time passes before half of a given quantity of the substance has been excreted.

Absorption, excretion, and conversion are especially thoroughly studied in the field of medicine, in which case knowledge is sought of wanted and unwanted effects in relation to the dose used.

3.6.1.2 Acute toxicity

Acute damages or "here-and-now effects" appear after short-term exposure to the substance.

In many cases, this is understood as the lethal dose expressed by the so-called LD_{50} value, but it is important to understand that a number of effects, such as headache and dizziness, may also be acute effects. This is illustrated by the following figures showing the effects of each dose.

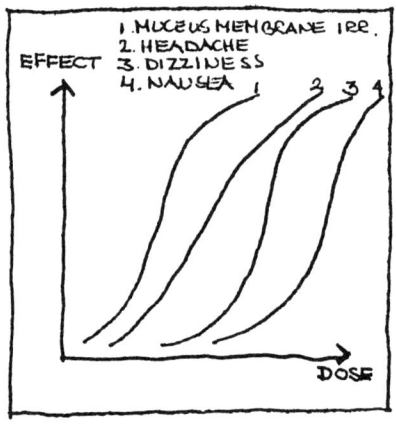
Dose-Effect Curve (21)

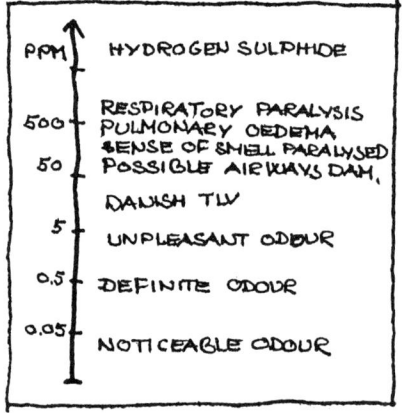
Correlation between dose and effect may also be illustrated this way.

The literature has a great many LD_{50} data determined by experiments on animals, usually rats. The animals are divided into groups, and to each group is administered a certain dose of the substance in question. The dose which kills 50% of the experimental animals in a group is called the LD_{50} value. The value most frequently used is LD_{50} oral (the substance is administered through the mouth); LD stands for "lethal dose" and is administered in milligrams of substance per kilogram of body weight.

LC_{50} is another value used: it implies that the substance is inhaled. (C stands for "concentration.")

It is important to seek information which describes effects other than death to human beings—if it exists. LD_{50} values do not tell us anything about the damaging or irritant effects a substance may give rise to acutely—such as headache, dizziness, slight illness, irritation of the respiratory passages, and the like.

In spite of this, only LD_{50} and LC_{50} values are on the whole used for the classification of substances and products according to the rules of the Danish Environmental Protection Agency (see Section 3.5)—and on the basis of this much too simplified system, substances and products with the highest values are preferred.

3.6.1.3 Skin and eye effects

Skin and eye effects in this connection are also acute (allergic eczemas are dealt with elsewhere—"Local irritation toxicology" is thus, to use the terminology of the Danish Environmental Protection Agency, for the classification and labeling of hazardous substances and products.

Some handbooks list graduated effects of chemical substances on skin and eyes in a number of categories, and most data are from experiments on animals.

The OECD and the EC (EC Directive 84/449/EEC) recommend that skin irritation be assessed in advance on rabbits, according to an exact procedure: the chemical substance is brought into contact with the skin for 4 hours, with a reading taking ½ hour after the period of contact; the skin is "read" as to irritation and formation of crusts, and each of these parameters is given a numerical value from 0 to 4. (The value 0 indicates no reddening irritation or formation of crusts.)

A number of animals are studied and the mentioned numerical values are added up to a single numerical value, a "skin irritation index."

Correspondingly, there are recommended tests for measurements of eye irritation.

These rules are also mentioned in the order of the Danish Environmental Protection Agency concerning labeling of hazardous substances and products as criteria of placement in the hazard classes "Corrosive" and "Irritant."

These tests are often carried out on new products with, e.g., substances which are new in the Danish market or if the manufacturer wants an "official" statement for his new product. One Danish company performs such tests, and there are a number of similar foreign companies.

3.6.1.4 Mutagenic effects

These are also called genetic damages and may lead to, among other things, miscarriages, birth defects, and cancer.

The genes are the information codes in all living cells. They determine the fate of the cells, e.g., which proteins are to be produced at certain points of time in the life of the cell. If the information codes are changed at an early stage, the result may be very serious.

If a gene is altered, it is called a mutation. If a gene in a cell is altered, the cell may die or alter its behavior. The cell may also live on as if nothing had happened. If the mutation takes place in a gamete (e.g., in an egg cell), the result of the mutation may be that the embryo of this cell is misshapen and perhaps the embryo ends as an early miscarriage. If the mutation takes place in a body cell, the result may be cancer.

A number of chemical substances are known to be capable of causing mutations in the genes of the cells. This is true of ultraviolet light and radioactive radiation, as well.

Direct studies are carried out on animals and microorganisms, e.g., bone marrow studies of mice, rats, and rabbits; studies of cultivated single cells from organs of various animal species; bacteria tests (e.g., Ames test).

Epidemiologic studies concern the observable consequences of mutations, namely, cancer, sterility, birth defects, etc.

3.6.1.5 Cancer

It is assumed that the type of cancer caused by most chemical substances is due to the genes of the cells being altered (the genes mutate). Then the cells behave in another way, that is, they split up into more parts beyond the normal body control of the behavior of the cells.

Some carcinogenic substances (carcinogens) are effective on the place of contact itself; others are transformed in the body so that the resulting products are in fact carcinogenic. Other carcinogens are only effective as initiators and require a cooperative substance in order to produce cancer.

There are a number of "official" criteria placing the chemical substances in groups which describe the certainty of carcinogenic effects. See, for example, IARC:[23]

1. Sufficient evidence of carcinogenicity in humans.

ASSESSMENT OF SUBSTANCES AND MATERIALS

2. Probability that the substance may produce cancer in human beings (limited data).

   2A. The agent is probably carcinogenic to humans, e.g., there is limited evidence of carcinogenicity in humans, and sufficient evidence of carcinogenicity in experimental animals.
   2B. The agent is possibly carcinogenic to humans, e.g., limited evidence in human, but absent of sufficient evidence in experimental animals.

3. Cannot be classified according to carcinogenic effect on human beings.
4. Is probably not carcinogenic in human beings.

In Reference 24, a number of substances are indicated which are placed in different lists, among others, the one of IARC. See also the list of carcinogenic substances at the back of the instructions of the Danish Working Environment Service relating to threshold limit values.[4]

The Danish Working Environment Service has made a strategy of action for the subject "occupational cancer."[25] See also Reference 26.

Studies of the carcinogenic properties of chemical substances may be short-term tests on microorganisms (bacteria, fungoid growths, single cells, etc.), long-term experiments on animals, and epidemiologic studies. The Ames test, which is performed on bacteria, is a typical microorganism test, often used as the first estimate of whether a substance is carcinogenic. The test deals only with the capability of the chemical substance to alter the genes of the bacteria. (As mentioned above, alterations of the genes are thought to be a possible mechanism of the development of cancer.)

As will be repeatedly pointed out in the following, it is worth noting the conditions of the experiments behind the toxicological data: When was the substance administered, what was the size of the dose, which species of animal was used, etc.?

3.6.1.6 Damage affecting reproduction

These include damage to egg cells, sperm cells, the embryo, and the child (by breast-feeding).

In the EC, a substance is only assessed as teratogenic if it causes congenital deformities. This criterion is too narrow, which is also pointed out in the strategy of action of the Danish Working Environment Service relating to substances damaging to reproduction.[27] The criteria are:

1. Reduced fertility
2. Damage to the genes
3. Damage to the embryo

The detailed criteria for how far a substance is considered capable of causing reproductive damage exists only for category 3, but international efforts are being made to define corresponding criteria for category 2.

Criteria for teratogenic substances have been established in this country (among others by the National Food Agency of Denmark and the National Institute of Occupational Health[28]). They may be summed up:[27]

Group 1 A chemical substance has teratogenic effects on human beings

- if congenital deformities have been proved in demographic investigations.

A chemical substance is considered to be teratogenic in human beings

- if congenital deformities have been proved in experiments with two species of animals or
- if congenital deformities have been proved in experiments with one species of animal and a very comprehensive knowledge of the substance exists in addition.

Group 2 A chemical substance is possibly teratogenic

- if congenital deformities have been proved in one species of animal and further knowledge about, e.g., the mechanism of effects of the substance exists in addition.

Data may be derived from both experiments on animals and from epidemiologic studies. However, relatively few substances have been studied as concerns effects damaging to reproduction.

There are several ways of examining whether a substance is damaging to reproduction.

A number of tests are performed on microorganisms. These tests show whether the genes are altered by introduction of the substance. An alteration of the genes in eggs and sperm cells of human beings may lead to damage to the embryo (miscarriage, congenital deformity, cancer, etc.); e.g., birth weight, resorptions, congenital deformities, number of dead/living embryos may be studied in experiments on animals, after the mother animals have been exposed to the substance. Sperm quality may be studied after the male animals have been exposed to the substance. Also, animal embryos in test tubes may be studied.

Studies of human beings may be epidemiologic studies of reduced fertility, congenital deformities, premature birth, reduced weight at birth, stillbirth, and infant mortality within, for example, certain occupational groups. In males, reduced number of sperm cells, abnormal or dead sperm cells, impotence, etc. are indicative.

It is important to note the experimental conditions: Is the substance administered before or after fertilization and for how long; to what quantities of the substance have the animals been exposed, etc.; is information available of the level of exposure in the epidemiologic studies?

Previously, there were no valid international lists of substances which are hazardous to reproduction such as those we have for, e.g., carcinogenic substances. The National Institute of Occupational Health therefore, in 1989, published a list with approximately 200 substances and groups of substances.[27] These substances are divided into groups according to three parameters:

1. Concentration level of effect
2. Type of effect
3. Level of documentation and quality of the studies

3.6.1.7 Allergy

This category comprises both allergic respiratory and skin diseases. As to skin allergies, tests may be used to reveal whether a substance causes allergy. It is a common feature of these tests that studies are divided into the following phases:

- Phase of administration (sensitization)
- Stage of rest (the allergy is allowed to develop)
- Stage of provocation (the substance is readministered and the reactions are studied)

One of the most frequent animal tests is the *guinea pig maximization test (the GPMT)*. The purpose of GPMT is to determine the risk of developing allergy in human beings.

The substance is injected under the skin of a group of guinea pigs, and 21 days afterwards the skin is tested by patch tests, i.e., the substance is "applied to the skin" under a special plaster. If a reaction occurs, it is probable that the substance also causes allergy in human beings.

If a person *has* developed allergy, the *cause* is often examined by means of *patch tests*. Thus, the purpose of this test is to find out whether a substance causes allergy in a particular person.

Other tests on animals include the Buehler test, open epicutaneous test, Draize test, optimizations test, Freund's complete adjuvant test, and the split adjuvant technique.[29]

The GPMT is among the most sensitive methods of assessing the allergenic properties of substances.[29]

Assessments of the applicability of the results of the mentioned methods can be found in the literature.[29-30] Experiences among human beings have been studied in epidemiologic investigations.

Studies of the capability of substances to produce allergies in the respiratory passages are not as standardized as experiments of skin allergies. They are often epidemiologic studies, but as is the case with the patch tests, persons may be tested for allergy in the respiratory passages by means of so-called provocative tests.

3.6.1.8 Other chronic damage to health

This includes damage to the lungs, liver, and brain.

Information has been obtained both from experiments on animals and from studies of human beings. Experiments on animals of this kind require, of course, more time and more resources, and the quantity of data is therefore not so extensive as that of data on acute damages.

Sometimes one comes across "exact" data corresponding to LD_{50} and LC_{50}, namely TDLo and NOEL, which mean, respectively, "toxic dose lowest" and "no observed effects level." These values are indicated in the same units as the LD values. It is important to note which effects have been examined.

ASSESSMENT OF SUBSTANCES AND MATERIALS 45

In some cases, relevant data may be obtained from studies of human beings who in their working lives have been exposed to the same chemical effects. These so-called epidemiologic studies comprise, for example, diseases and specific damages to the organs in a specific occupational group or people in a special factory. Comparisons are made with a control group which has not been exposed to the chemical effects. Causes of death are studied as well.

In epidemiologic studies it is difficult to determine the exact dose and the exact time of exposure. The connection between a concrete exposure during the working life (defined as well as possible) and the frequency of a specific disease in the exposed occupational group is analyzed. If such a connection reveals a relation between cause and effect, it depends on the availability of information about possible competing causes of the disease in the persons' lives. It is, thus, important to know about tobacco smoking when the connection between cancer in the lungs and exposure to asbestos is studied.

Often is merely a statistically reliable connection between the profession and one or more diseases, perhaps also dependent on the number of years of occupation in the profession.

3.6.2 Substitution by Means of Toxicity Assessment

Contrary to the previously mentioned tools, numerical values may seldom be compared directly with a toxicity assessment to decide if a substitution is successful or not.

It is often necessary to compare original data of experiments performed under different conditions. Some methods assign numbers to the toxicity (LD_{50} values, skin and eye irritation index, etc.).

Some of the sources used for toxicity assessments, especially various handbooks, working environment encyclopedias, etc. have already assessed the mentioned original data and pass on the results in the form of numerical values on a scale, or as statements such as, e.g., "may cause damage to the liver," "produces cancer in mice and rats," "under suspicion of causing damage to the liver." The bibliography includes the reference works that the Danish Environmental Protection Agency requires as a minimum for the assessment of the hazard of a chemical substance.

This means at the same time that the data to be compared may have been interpreted in different ways, dependent on the views of the interpreter. It is therefore important to be familiar with the conditions of the experiments.

It may be difficult to keep control of the extensive and varied spectrum of toxicological data which have been collected. It is therefore important to systematize the treatment of the data. First and foremost, the component substances are assessed one by one. The table at the end of the section about toxicity assessment may be useful in this process: one table is filled in for the component substances of each product. Finally, one may try to fill in a total table of the product.

In L. Seedorff's book *Substitution* (see the general part of the bibliography) toxicity assessment is of considerable importance in questions of substitution.

3.6.3 Example

We have not made a thorough toxicity assessment of the example of the PVC adhesives. The information on the following page is from Sax,[31] which is a much used work of reference.

It is clear that the LD_{50} values are not good indicators of the hazardous effects of PVC adhesives. Furthermore, the table shows that product 1 contains substances which may have an effect on the nervous system and cause birth defects. Moreover, liver and kidneys may be damaged. Product 2 does not have these effects on health, according to the reference work used. On the basis of these data, product 2 would therefore be preferred.

3.6.4 Discussion

The toxicity assessment cannot be reduced to a comparison of numbers—such as, e.g., LD_{50} values and skin irritation indications. In each case the information procured has to be compared and assessed, which is often quite difficult. For example, how is a skin-irritating product assessed in relation to a product which may cause permanent damage to the brain? It is therefore important to know the working process exactly: if there is no risk of contact of the product with the skin, the assessment may be facilitated. One condition of making a good toxicity assessment at all is that the product to be assessed is described by means of a complete declaration of content, which is difficult to obtain, as many Occupational Health Service workers know. Is it, e.g., necessary to require information about components in the product below a certain quantity (e.g., 0.1%)? And what about impurities? Carcinogenic and allergenic substances may often damage health even if the quantity in the product is small.

Nevertheless, reasonable declarations of content are often obtained from the suppliers—declarations which also inform of the components in small quantities. But these declarations of content are frequently made available on condition of a vow of secrecy, and to maintain the reputation of trustworthiness of the Occupational Health Service it is important that this is respected.

ASSESSMENT OF SUBSTANCES AND MATERIALS

Substance	LD_{50}, oral, rat	Toxicological information[a]	Code[a]
THF	3,000 mg/kg	Irritates eyes and mucous membranes. Moderate toxic effect by oral administration and by inhalation. May damage liver and kidneys.	2
DMF	2,800 mg/kg	Affects the human central nervous system. Teratogenic in experiments on animals. Moderate to low toxic effect in a number of animal species.	3
MEK	3,400 mg/kg	Moderate toxic effect by consumption. Low toxic effect by absorption through the skin. Extreme irritation (of eyes and mucous membranes). Affects the peripheral nervous system and the central nervous system. Teratogenic in animal experiments.	3
NMP	4,200 mg/kg	Moderate toxic effect by direct absorption through the blood (intravenous) and from the abdominal cavity. Low toxic effect by oral administration or absorption through the skin.	2 or 1

[a] *In this reference, a code has been assigned to the substances (there are five possible: U = no data, unknown toxic effect, 0 = no toxic effect, 1 = low toxic effect, 2 = moderate toxic effect, 3 = high toxic effect). The assignment of code is performed under consideration of both acute local effects and effects on organs and chronic local effects and effects on organs. Such a division of chemical substances may be useful in a toxicity assessment but may be problematic as well. How much are, e.g., long-term effects weighted in relation to acute skin and eye effects?*

If exhaustive information about the content does not exist or cannot be obtained, it may be possible to procure an assessment of the products from the Register of Substances and Materials of the Danish Working Environment Service, but one must be prepared to wait for a long time. Perhaps the "Chemical Service" of the Danish Working Environment Service may be of help.

A detailed toxicity assessment takes time. It should therefore be considered when it is necessary to make one's *own* assessments and when those of other workers can be used—and who of those others may be trusted. Thus it is also necessary to realize at what level the toxicity assessment shall be performed. How far is the matter to be dealt with in detail, before the assessment of the product is justifiable?

Moreover, there is often a lack of toxicological data (especially in the case of new substances) concerning long-term effects, damage to reproduction, cancer, etc.

It is important to study the conditions of the experiments—which species of animals are used, which doses have been administered to them (and for how long). In epidemiologic studies it is important to know the levels of exposure of the persons, age, smoking and alcohol habits, etc.

It is difficult to convert data from animal experiments to human beings. It is, for example, not certain that LD_{50} values, which are indicated in mg/kg body weight, can be converted to human size by multiplication of the weight of the human being. There are suggestions, e.g., that the surface area of the animal/human being is of greater importance than the body weight. In that case, the human being is 6 to 7 times more sensitive than rats and approximately twice as sensitive as dogs. Diethylene glycol has a LD_{50} value of 14,800 mg/kg in the case of rats; for human beings the same value is approximately 1000 mg/kg. The age of the animal/human being is also important: sodium nitrite has a LDLo value of 330 mg/kg for dogs; the same value is approximately 22 mg/kg for children.

Even if size and weight are not essentially different, metabolism (conversion and excretion) may be different, that is, rats do not convert aromatic amines in the same way as do human beings and rabbits, which influences the estimate of experiments with these substances on rats.

3.7 Current Norms

As the heading indicates, this tool is not as systematic as, e.g., TLV, VHR, MAL coding, and toxicity assessment. However, we have chosen to include "current norms" as a tool, because we think it is much used, and often with considerable success. What we mean by "current norms" may best be described by a number of examples:

ASSESSMENT OF SUBSTANCES AND MATERIALS

Collection of data for toxicity assessment.

Type of toxicity	No data	Biological system (rat, mouse, etc.)	Type, degree, volume of toxicity	Experimental conditions (e.g., quantity, time and method of exposure)	Reference
Acute toxicity					
Toxicity after long-term exposure					
Toxicity of reproduction					
Toxicity of the genes					
Cancer					
Allergy					
Local irritation (skin and eyes)					
Miscellaneous					

Ref.: MalmforsConsult AB, Stockholm.

- Change of solvent-based products to water-based products, e.g., within the categories of paints, cleaning and degreasing agents, and coolants/lubricants.
- Change of products with short-chain hydrocarbons to products with long-chain hydrocarbons.
- Substitution avoiding a number of groups of chemical substances, e.g., epoxy, amines, lead and other heavy metals, asbestos, and others.

Current norms often arise from thorough studies, but are often simplified in their final version, so that differences are blurred. Under some conditions the substitution functions well, but in connection with certain special working processes, perhaps dependent on the physical conditions in the workroom, it functions badly.

Thus, if there are hot surfaces nearby the workplace, or if welding is performed, it is perhaps important to examine whether the components of the new product can be converted into toxic gasses when heated or exposed to ultra-violet light. This may be relevant if aromatic solvents are replaced by chloric-fluorine ones, since these can be converted to, e.g., phosgene and phosgene-like substances under such conditions.

Even if the substitution is successful, it must be remembered that the new product probably is not *nonhazardous* but is *less hazardous,* and thus should be handled with care.

We shall not discuss this tool further here, but only point out that substitution should not be carried out uncritically according to "current norms."

3.8 The Experiences of Other Workers

One of the major purposes of this book is to make the substitution experiences of other persons accessible to a wider circle. It is one thing to make the necessary assessments oneself, but another thing to exchange experiences with persons who have assessed, considered, substituted, and—not least— have seen the substitution carried out in practice and received feedback from the workers. Making the same or a similar substitution in another company requires, of course, separate assessments, but no doubt the experiences of other persons can be used positively for these assessments. Reference is made to Part 2 for a further discussion of substitutions in the Occupational Health Service.

ASSESSMENT OF SUBSTANCES AND MATERIALS

3.9 Miscellaneous

Apart from the above-mentioned tools, there are a number of other more or less specific tools for use in the assessment of substances and products. The following list is not exhaustive; among other things, a number of tools for the assessment of technical properties are not included, properties which are often decisive if a substitution is to be successful.

3.9.1 Classification of Welding Electrodes

Coated electrodes for welding purposes are classified according to order no. 407 from 1979,[32] issued by the Ministry of Labour. The classification relates to the smoke that comes from welding. There are seven classes; class 1 develops the least smoke, class 7 the most.

The importer/manufacturer is responsible that the electrodes are provided with the correct smoke-class number. The classification must be carried out by an authorized test institute.

For purposes of substitution, the lowest class is preferable.

3.9.2 Measurements of Chemical Substances in the Air

Instead of calculating the risk of exceeding the threshold limit values (by means of, e.g., TLV and SUBFAC), the air may be measured. In the example of the PVC adhesives, the replacement of product 1 with 2 was motivated by such measurements.

In this connection, measurements in the workplace give a more realistic impression of the conditions than do laboratory measurements and calculations, but they are subject to greater uncertainty because of variations in the performance of the work, thermal conditions, etc.

In the example mentioned, the calculations would provide the same result and the measurements would not be necessary.

3.9.3 Smell

Whether a chemical substance smells good or bad and in which concentrations of air these qualities of odor appear are subjective conditions differing from the other tools that have been described. A substitution may be motivated by the ill or irritant odor of a substance. However, the "quality" of the smell and the limits of smell are not immediately connected with the hazardous properties of the substances, e.g., carbon monoxide cannot be smelled even in concentrations far above the threshold limit value, whereas the smell limit of hydrogen sulfide is far below the threshold limit value. Moreover, the latter substance blunts the sense of smell so much that the smell limit is soon raised.

Reference 33 describes the quality of smell and indicates smell limits and irritation limits of 450 chemical substances. Smell limits are also found in Reference 34.

As mentioned above, smell may motivate substitution but smell cannot be used as a tool of substitution.

3.9.4 Material Safety Data Sheets

Material safety data sheets are often used as a basis of substitution. The rules for the content of instructions of a chemical substance or material are described in Reference 1. Here it says, among other things, that data sheets shall provide information about

- Trade name
- Applicability
- Limitations of use

ASSESSMENT OF SUBSTANCES AND MATERIALS

- Requirements of special training
- Properties hazardous to health
- Measures to be taken when using the substance or material
- First aid
- Properties when exposed to heat or fire
- Measures to be taken in case of fire
- Measures to be taken in case of spillage or removal
- Safety regulations for storage
- Classification and labeling

This information should provide a good background for the assessment of chemical products. But safety data sheets are often of a very differing quality. Some instructions describe in detail the hazardous properties of the substances, whereas others scarcely do so. Besides, safety data sheets must be updated when new information about the hazardous properties of the substances appear, which is not always done.

Some professional groups have made preprinted material standard safety data sheets for certain groups of products; then the suppliers are to fill in the relevant information. This is the case with, for example, the association of Danish paint and dye manufacturers and the association of Danish adhesive manufacturers.

At the same time this results in a homogeneous quality level of the instructions. The quality of instructions is discussed in Reference 35.

Part II

4. THE STUDY

4.1 Background of the Study

In recent years much substitution work has been carried out in the Danish Occupational Health Centres. The experiences have hitherto been communicated sporadically through telephone contact and more systematically at the yearly exchange of experiences in the Occupational Health Service seminars.

To continue the substitution work, it is important to build on the experiences which already exist—for better or for worse.

A group of workers of the Occupational Health Service have therefore, since the 1986 seminar, worked at collecting the knowledge of substitution that exists in the Occupational Health Services and making this knowledge widely accessible to the workers of the Service. Collection of these experiences has been made by means of a written poll.

4.2 Purpose

Why this study?   The purpose of the poll was to collect concrete examples of substitutions made by the Occupational Health Service. Furthermore, the purpose was to communicate these examples in a form which may serve as inspiration and a store of knowledge to the benefit of other Occupational Health Service staff.

As an extra bonus, the results gave us some ideas about the substitution work in the Occupational Health Service generally.

4.3 Method

How was the poll performed?   Three questionnaires and an enclosed letter (see Supplement) were sent to all Occupational Health Services in Denmark according to the mailing list of the Danish Working Environment Fund. This material was sent to 127 Occupational Health Services of all types. If a trade Occupational Health Service has a principal department and several geographically spread subdepartments, it was only sent to the principal department.

Schedule

Sent out	20-07-87
Response time limit	06-08-87
Written reminder	12-08-87
Last response received	02-10-87

When all examples had been entered and processed, they were returned to the person dealing with the matter with an accompanying letter, giving a further opportunity to express an opinion about the information to be disseminated. See the reading list in the example section.

4.4 Material

How many Occupational Health Centres answered?

Number of answers and percentage of answers appears in Table 4-1 below.

Some answered without examples

There were answers from 10 Occupational Health Services without examples of substitution. We have included these 10 in the answers. Of the 10, 5 answered that they were very positive about the initiative but that they had not yet had practical experiences of substitution. A company Occupational Health Service answered that, according to management, there were no experiences of substitution.

How many filled-in questionnaires did we receive?

This part of the report is based on 162 questionnaires filled in with examples of substitution. See Table 4-2.

Table 4-1.

Letters mailed	127
Answers from Occupational Health Centres	53
Percentage of answers	42%

Table 4-2.

Number of returned questionnaires	168
Number of discarded questionnaires	6
Number of questionnaires about substitution	162

The work group has, as previously described, had the following conception of substitution as its starting point:

The background of substitution is the presence of a hazardous substance or material. In order to substitute, the chemical substance or material which is the cause of the considerations of substitution must be removed from the working process. This may take place on one of the three levels. See Chapter 1.

Some were discarded — On this background, six of the returned questionnaires were discarded, since we at this time did not want to deal with situations in which the risk of exposure to a hazardous substance/material is removed by an exhaust system or other technical measures. Examples are not included in which the cause of the change is other than chemical, e.g., a high noise level.

4.5 Quality of Data

The questionnaires are well filled in — The 162 examples of which the material consists are well described. People have made efforts to describe their experiences within the framework of the questionnaire.

The contents are people's own answers and their own responsibility — Because the answers describe concrete situations, we did not engage in an estimate of the contents of each answer. Thus the quality of the performed substitution has not been estimated (i.e., whether a real improvement of the working environment has taken place). Our reports and count comprise people's answers to the questions. The evaluation is left to the reader.

Vow of confidentiality — The Occupational Health Service has a vow of confidentiality. In Section 5.1 the vow is described.

Representative character? — The purpose of the poll was not to obtain a representative survey of substitution work in the Occupational Health Service, but to collect as many examples as possible. On the contrary, we are almost sure that our material is not representative. 58% of the Occupational Health Services in the country have not answered, and we do not know what takes place there. Moreover, each Occupational Health Service worker has probably sorted out according to experience what he wanted to pass on. We did so ourselves, in any case. Thus the results of this report cannot be generalized.

4.6 Results

4.6.1 Types of Products—What Do We Do?

The Occupational Health Service often substitutes solvents

As appears from the examples (Chapter 5), most of the substitutions replace a solvent or a compound of solvents with other chemicals, which may for that reason be other solvents.

In Table 4-3, selected organic solvents are listed with vapor pressure, VHR, and the threshold limit values of the substances.

Table 4-4 shows selected substitutions for solvents. The symbols indicate which solvents are changed from and to what. The solvents are listed according to decreasing VHR. VHR is obviously often—consciously or unconsciously—used for the decision of substitution. There is a clear tendency that substances with high VHRs, i.e., with low threshold limit values and/or high vapor pressures, are replaced by substances with considerably lower VHR.

Cleaning/degreasing, adhesives, colors

Among the types of products most often involved in substitutions, most frequent are cleaning agents (a total of 53), followed by adhesives (27), colors and paints (15), and coolants/lubricants (10). The other types of products are represented by 7 or less.

Table 4-3. The Vapor Pressures, Threshold Limit Values (TLV) and Vapor Hazard Ratios (VHR) of Selected Substances

Substance	Vapor pressure ppm/°C	TLV 1988 ppm	VHR
Chloroform	207,000/20 (36)	2	103,000
Benzene	125,000/25 (37)	5	25,000
Tetrachloromethane	20,800/22 (38)	2	10,400
Methylene chloride	500,000/22 (38)	50	10,000
n-Hexane	197,000/25 (39)	50	3,940
Trichloroethylene	79,000/20 (36)	30	2,633
Acetic acid	15,000/20 (38)	10	1,500
1,1,1-Trichloroethane	132,000/20 (38)	100	1,320
Butanone	131,000/25 (40)	100	1,310

Ethylene glycol	5,000/20 (38)	5	1,000
Acetone	237,000/20 (36)	250	950
Tetrahydrofuran	170,000/20 (36)	200	850
Toluene	36,800/25 (41)	50	736
1,1,2-Trichloro-1,2,2-trifluoroethane	355,000/20 (36)	500	710
Methanol	132,000/21 (38)	200	660
Formic acid	2,770/24 (38)	5	554
Ethyl acetate	101,000/20 (36)	300	337
2-Propanol	57,800/25 (42)	200	289
Cyclohexanone	6,888/25 (43)	25	272
Isophorone	1,320/38 (38)	5	264
Ethanolamine	632/20 (36)	3	211
Xylene	8,840/21 (38)	50	177
Heptane	52,600/22 (38)	400	132
Solvent naphta (as heptane)	52,600/22 (38)	400	132
Propylene glycol methyl ether	10,500/20 (44)	100	105
Phenol	462/25 (45)	5	92
Propanol	13,200/15 (38)	200	66
Ethanol	52,600/19 (38)	1000	53
Odorless petroleum	5,000/25 (46)	100	50
White spirit	5,000/25 (47)	100	50
C(8-14)-aromatics	500/20 (44)	25	20
N-Methyl-pyrrolidone	525/20 (44)	100	5
Ethylene glycol	128/25 (48)	50	3
Propylene glycol	104/25 (49)	—	—
Water	/	—	—
Other liquids with very low vapor pressure, solid substances or no substances	/	—	—

Satisfaction with the substitution

In the following we will attempt to gather the threads of each product group to make a subjective proposal as to what is likely to succeed within substitution. Only product groups addressed by at least three questionnaires have been included.

Antifreezes   Replacement of ethylene glycol by propylene glycol as coolant usually causes satisfaction; propylene glycol may, however, irritate the mucous membranes. See miscellaneous about denatured ethanol.

THE STUDY

SUBSTANCE	VHR
Choloroform	103,000
Benzene	25,000
Tetrachloromethane	10,400
Dichloromethane	10,000
n-Hexane	3,940
Trichloroethylene	2,633
Acetic acid	1,500
1, 1, 1-trichloroethane	1,320
Butanone	1,310
Ethyl glycol	1,000
Acetone	950
Tetrahydrofuran	850
Toluene	736
1, 1, 2-trichloro-1, 2, 3-trifluoroethane	710
Methanol	657
Formic acid	554
Ethyl acetate	337
2-propanol	289
Cyclohexanone	272
Isophorone	264
Ethanolamine	211
Xylene	177
Heptane	132
Solvent naphta (as heptane)	132
Propylene glycol methyl ether	105
Phenol	92
Propanol	66
Ethanol	53
Odourless petroleum	50
White spirit	50
C(8-14)-aromatics	20
N-methyl pyrrolidone	5
Ethylene glycol	3
Propylene glycol	-
Water	-
Other fluids with extremely low vapor pressure, solid substances or no substances	-

o: "to" substance   ●: "from" substance   ▲: both "to" and "from" substance

Paint removers	Replacement of paint removers containing methylene chloride with alkali, scouring powder, etc. has been successful in a couple of cases. On the other hand, a change of procedure from immersion in products with, e.g., phenol for sandblasting has not been successful in a couple of cases because of financial or ergonomic problems. No change was made from the use of a hand cleanser to a spray product with *N*-methyl-pyrrolidone. A mixture of 2-butanone and acetone was changed to pure acetone with a good technical result (but from the point of view of health, it could have been done better, the Occupational Health Service worker meant).
Color and paint	This group is satisfied with the performed substitutions, which as a whole are as follows: • reduced quantity of (volatile) solvents • removal of lead and chromate • lower content of low ethylene glycol ether • reduced degassing of formaldehyde
Coolants and lubricants	There is satisfaction with the performed substitutions, which can apparently be carried out with almost all component substances. However, there is dissatisfaction with problems of stability in storage and uncertainty about other component substances.
Laboratory chemicals	In most cases, there is satisfaction with the replacement of carcinogenic solvents, e.g., chloroform and benzene by "common" solvents. In one case, the substitution did not function technically satisfactorily. In another case, a powdery carcinogenic substance was replaced by a tablet of the substance without problems.
Adhesive	In several cases, adhesives have been replaced by mechanical or physical methods with a satisfactory result. Besides, many adhesives with, for example, methylene chloride, cyclohexanone, and/or tetrahydrofuran have been replaced by adhesives containing among other things *N*-methyl-pyrrolidone, as a rule with a satisfactory result; however, the health properties of the latter substance are questioned. Besides, there are several ex-

THE STUDY

amples of shifts between other solvents or solvents and water, usually with a satisfactory result.

Cleaning agents — There is usually a satisfactory result of the performed substitutions, in which the tendency of solvents is almost as indicated in Table 4-4. However, there are also many examples of changes to water-based solvents.

The tendency of degreasing agents is like that among the cleaning agents (see above). Many workers in this group have expressed their satisfaction with the substitution.

Releasing agents — The replacement of releasing agents based on mineral oil with water-based or vegetable oil-based products is as a rule successful. There may be problems of dust from cement after suspension in water.

Lubricants — A rinsing agent could be entirely removed and silicone could be replaced by soft soap without problems. Petroleum could be replaced by nonvolatile white spirit with a satisfactory result.

Fillers — No tendencies in this group.

Miscellaneous — This is also a very mixed group without common tendencies. As mentioned previously, methylated spirit is dealt with here. Ethanol denatured by pyridine has in three cases been successfully replaced by 2-propanol prepared ethanol. Change from this product to another with vegetable turpentine also caused satisfaction. In one case, ethanol was denatured with pyridine and amyl alcohol changed to no content of pyridine, but the workers were still concerned about the health hazard.

In Section 5.2, a survey of the performed substitutions classified according to product groups is presented.

4.6.2 Who Benefits from the Substitutions/Is Exposed to the Substitutions?

Approximately 40% of the substitutions concerned unskilled workers within different professional groups.

Professional group	Number of questionnaires
Unskilled	64
Skilled, building/construction	21
Smiths	5
Mechanics	6
Engine fitters	6
Laboratory assistants	6
Skilled + unskilled, others	38
Unexplained	16
Total	162

4.6.3 Why Do We Set About Substituting?

There may be many different reasons for setting about substituting. By far the most frequently mentioned cause in the answers was fear of possible damage to health on the basis of a toxicity assessment of the work at the product (72 answers) or because the workers had symptoms or were encumbered with irritant effects (62 answers).

Task on the background of symptoms

Thus, 62 of the tasks performed had a background in registered symptoms and irritant effects. Especially frequent were headache, nausea, and slight illness in connection with work with organic solvents. Skin effects and smell irritations often occurred, too.

Of these tasks, two thirds had been carried out to the satisfaction of the caseworkers. In half of the cases they had succeeded in making the symptoms and the irritant effects disappear entirely or at least reducing them considerably. The rest of the satisfied persons judged the result from toxicity assessments or measurements of pollution. Dissatisfaction prevailed if the new product:

- for various reasons was not used
- could not satisfy the technical requirements
- also had unknown or problematic properties hazardous to health
- had given rise to new irritant effects or symptoms.

Other reasons for initiating tasks

The following items were mentioned as reasons for initiating substitution tasks:

- measurements (7)
- reorganization of working processes (8)
- order from the Danish Working Environment Service (7)
- description of the products used (3)
- economy (1)

On the basis of the questionnaires, it does not seem that the background of the tasks has essentially influenced the satisfaction. However, it must be mentioned that few are satisfied with solutions which are a result of having to comply with ministerial order no. 540. Nor does the substitution initiated with a background of financial interest encourage beginning experiments of substitution.

4.6.4 Which Methods Do We Use?

Which tools are used for substitutions? In Table 4-5, the most important methods are indicated. More than one method could be indicated in each questionnaire. 188 answers from 162 questionnaires are recorded. The distribution is from 0 to 4 per questionnaire.

It appears from the table that toxicity assessment is the most frequent method of substitution. It is impossible to judge the level in which these assessments have been made.

However, there is no doubt that VHRs or related methods—MAL coding and threshold limit values—are essential to the decision of what is replaced and with what.

Table 4-5. Methods used for Substitution. The Methods Marked with * were Indicated as Possibilities in the Questionnaire.

		Number (% of questionnaires)
*	Toxicity assessment	79 (49%)
*	Threshold limit values	17 (10%)
*	MAL code	14 (9%)
*	Vapor hazard index	14 (9%)
*	Experiences of others	11 (7%)
	Smell	6 (4%)
	Literature	4 (2%)
	Background knowledge	3 (2%)
	Technical requirements	3 (2%)
	Own experiences	3 (2%)
*	Unsympathetic supplier	3 (2%)
	Others (including * labeling (2)) and * SUBFAC (1)	31 (19%)

Nobody mentions directly that substitution is carried out on the basis of classification and labeling of the products.

A few mention "unsympathetic supplier" as a cause. This means that the choice between two products is influenced by the cooperation of the supplier. Is it easy to get information? Are experiments favored?

4.7 Conclusion

It is necessary that the experiences we get in the Occupational Health Services are picked up and coordinated. We have tried to do so in the case of substitution. Against this background some questions arise:

1. How shall we collect and coordinate the experiences and examples of substitution in the future?

2. Not only other Occupational Health Services may profit from experiences and examples of substitution. It is especially important to communicate the idea of substitution if it is general in character and can be directly transferred to other companies. How do we communicate more generally? Through which channels? What does the Danish Working Environment Service think?

3. Substitution is only one of many areas which require a nationwide coordination in the Occupational Health Service. How do we meet this need of coordination so that the solution has a more general and permanent character?

5. EXAMPLES

5.1 Reading Instructions

In this chapter, 162 examples of substitution tasks performed by Occupational Health Service workers are indicated. The examples are taken from a questionnaire study among all the Occupational Health Centres in the country in August 1987. The method and material of the study are described in Sections 4.3 and 4.4.

Index of product groups	The examples are arranged according to product groups so that products with the same function are gathered in one place. "Journalisering I arbejdsmedicinen" (J.P. Johansen) has been used for systematization. An index of product groups is given in Section 5.2. This index may

	be used as a primary entry to the examples, if experiences with replacement of coolants/lubricants (for example) are sought.
The examples	In Section 5.3 the examples are listed according to the product group index, Section 5.2. In the examples we have chosen to reproduce all answers in their complete, original wording, because we think that the examples in this way may function optimally as a catalogue of ideas. Both good and bad experiences are included, from the consideration that it is very important to avoid the repetition of bad experiences in practice.
Catalogue of ideas	The intention is that this catalogue of ideas should function as an inspiration to contact Occupational Health Service workers experienced in substitutions in which you are interested.
No answers given beforehand	It is not intended that the examples shall be directly used for "go and do likewise." Reality is complex and the questionnaires do not comprise all reservations and descriptions of more or less mystical conditions which it may be necessary to know.
Index of professional groups	It is also possible to seek experiences of substitution within specific professional groups. In the professional group index, Section 5.5, all substitutions are listed by number under the professional group in which the substitution has been performed. Since the examples are numbered consecutively, it is easy to find the examples. If one wants to know about experiences in plumbing, for instance, the index indicates which examples are from that professional group.
Vow of confidentiality	As mentioned above, the Occupational Health Service is under a vow of confidentiality according to Occupational Health Service order §18. The vow of confidentiality concerns the knowledge the Occupational Health Service workers acquire through their work. This means in practice that it must not be possible from the descriptions to find out which company is described and that product secrets must not be revealed.

EXAMPLES

When designing the questionnaire we asked the Occupational Health Centres to answer with examples which may later be part of a joint database.

Examples returned to the senders

After registration and further work with the material, the examples of each Occupational Health Service have been returned to the case workers. We wanted thereby to give them another chance of expressing their opinion on the following:

1. Whether there are instances when the vow of confidentiality of the Occupational Health Service has been broken. In the original questionnaire we had, as mentioned above, pointed out that the information should not be confidential but should, if necessary, be passed on.

2. Whether some information should not be included in consideration for suppliers.

3. Whether misunderstandings owing to misprints had arisen in the process of registration.

The answers were to be returned within a fortnight. We pointed out in the enclosed letter that failure to answer was considered as an approval.

Misspellings

We received in all 10 replies to this request. Most of them dealt with errors of writing or misspellings. Some had obtained new information concerning substitution after the return of the questionnaires. These are *not* added. The work of substitution in the Occupational Health Service is a process having new events all the time. The examples must therefore be considered as a state of affairs per September 1987—and much new may fortunately have occurred since then.

Cooperation with suppliers

Finally, some talked less kindly of a couple of suppliers, but they did not want their statements printed because of the cooperation between the Occupational Health Service and the suppliers. We have of course corrected it, but they still appear in the counts.

5.2 Survey

Substitutions classified according to product groups
(Short summaries)

FROM	TO

ANTIFREEZES
001 Ethylene glycol — propylene glycol
002 Ethylene glycol — propylene glycol
003 Ethanol (pyridine/amyl alcohol) — containing propylene glycol
004 Containing propylene glycol (smell) — ethanol (2-propanol)
005 Ethanol (pyridine)

FUELS
006 Petrol/diesel — electricity

PAINT REMOVERS
007 Methylene chloride/methanol — scouring powder/hand cleanser/paint
008 Methylene chloride/methanol (5-3) — alkali (00-4)
009 Acetone/methyl ethyl ketone (87:13) — acetone
010 Ethanolamine/phenol — sandblasting
011 Hand cleanser — N-methyl-pyrrolidone (spray)
012 o-Dichlorobenzene/phenol — sandblasting

COLORS AND PAINTS
013 White spirit/2-propanol — propylene glycol/Texanol 00-1
014 Woodwork paint 2-1 — woodwork paint 0-1
015 Solvents incl. toluene — solvents excl. toluene
016 Solvent-borne and latex paint — silicate paint
017 Common silk screen printing colors — aqueous silk screen printing colors
018 Lacquer with lead chromate — lacquer without lead chromate
019 Lead/chromate/solvents — smaller quantity of solvents without lead/chromate
020 Lead/chromate 4-3 — without lead/chromate 4-1
021 Lead/chromium — without lead/chromium

EXAMPLES

022 Ethyl glycol/methylglycol   without ethyl
                                       glycol/methylglycol
023 UV-lacquer/formaldehyde,   formaldehyde, low evaporation
    high evaporation
024 Aliphatics/white spirit/toluene   aliphatics/propanol
025 Ethyl glycol/ethanol   ethanol/water
026 Oil-based stencil paint   water-based stencil paint
027 Preservative (boric acid)   without preservative

DYES
028 Dyeing powder   aqueous dispersion

DILUENTS
029 Xylene   white spirit
030 2-Propanol   ethanol (denatured by 2-propanol)

PRECIPITANTS
031 Flocculant (hydrocarbons)   flocculant without hydrocarbons

SKIN CLEANSERS
032 With organic solvents   gloves/cream/soap
033 Lanolin/citrus oils   ?

IMPREGNATING AGENTS
034 Thread with formaldehyde   other thread
    releaser/siloxane/oil/
    paraffin

INSULATION MATERIALS (heat, etc.)
035 Insulation thread with   insulation thread without asbestos
    asbestos

INSULATION MATERIALS (electricity, etc.)
036 1,1,2-Trichloro-1,2,2-   oil
    trifluoroethane

PRESERVATIVES
037 Acetic acid   citric acid

COOLANTS AND LUBRICANTS FOR METAL WORK
038 1,1,1-trichloroethane   rape-seed oil
039 Mineral oil/nitrite/amines/   water/amines/without
    formaldehyde   formaldehyde

040 Mineral oil — synthetic and half-synthetic coolants/lubricants
041 Coolants/lubricants with chlorinated paraffin — coolants/lubricants without chlorinated paraffin
042 Petroleum sulfonate/EP-additive — coconut fatty acid/boric acid complex
043 Chlorinated paraffin — without chlorinated paraffin
044 With biocide — without biocide
045 Coolants/lubricants dissolving cobalt — not cobalt dissolving
046 With nitrite — without nitrite
047 Common coolants/lubricants — with antiphosphine additive

LABORATORY CHEMICALS
048 Benzene/2-propanol — toluene/2-propanol
049 Tetrachloromethane/chloroform — esters/ketones
050 Chloroform/acetone — ethanol/ethyl acetate/n-heptane
051 Chloroform/methanol/formic acid — toluene/acetone/acetic acid
052 o-Phenylene diamine powder — tablets

ADHESIVE
053 Methylene chloride/tetrahydrofuran — mechanical joining
054 Methylene chloride — heat
055 Tetrahydrofuran/cyclohexanone — mechanical joining/joint filler
056 Methylene chloride — joining by sleeves
057 Solvent — water
058 1,1,1-Trichloroethane/acetone/toluene — water
059 Solvent naphtha (n-hexane) — water
060 Solvent — water/heat
061 Polyurethane (MDI)/water — solvent
062 Organic solvent — without organic solvent
063 Organic solvent (among others, toluene) — hot-melt
064 Hot-melt advanced opening time — hot-melt retarded opening time
065 Hot-melt — self-adhesive tape
066 Chloroform — ethyl acetate
067 Solvent — aqueous/30% toluene

EXAMPLES

068 Methylene chloride   *N*-methyl-pyrrolidone
069 Methylene chloride   *N*-methyl-pyrrolidone
070 Solvent   *N*-methyl-pyrrolidone
071 Methylene chloride/   *N*-methyl-pyrrolidone
    tetrahydrofuran
072 Tetrahydrofuran/   *N*-methyl-pyrrolidone
    cyclohexanone/butanone
073 Tetrahydrofuran/   *N*-methyl-pyrrolidone
    cyclohexanone/methylene
    chloride
074 Tetrahydrofuran/   *N*-methyl-pyrrolidone
    cyclohexanone
075 Butanone   *N*-methyl-pyrrolidone
076 Butanone   *N*-methyl-pyrrolidone
077 Heptane/toluene/ethyl acetate   heptane/toluene/ethanol
078 Epoxy   method indicated
079 Asphalt with adhesive   without
    enhancer

SOLDERING PRODUCTS
080 Lead/colophony/organic   silver solder/water-washable flux
    solvents

CLEANING AGENTS
081 Polymethylumcrylate   abrasive/mineral oil, etc.
082 Trichloroethane   alkali, etc.
083 1,1,2-Trichloro-1,2,2-   alkali
    trifluoroethane
084 1,1,2-Trichloro-1,2,2-   alkali
    trifluoroethane
085 1,1,2-Trichloro-1,2,2-   "Ajax" for windows
    trifluoroethane
086 2-Propanol/water   propylene glycol/water
087 (C_{11} to C_{12}) *n*-alkanes   alkali
088 White spirit   alkali
089 Petroleum   alkali
090 Alkali   warm soap solution
091 Ethanol(den)   water-based cleaning agent
092 Methylene chloride   toluene
093 Methylene chloride   *N*-methyl-pyrrolidone, etc.
094 Chloroform   ethanol
095 1,1,1-Trichloroethane   *N*-methyl-pyrrolidone or alkali
096 1,1,1-Trichloroethane   glycol ether

097 Ethyl glycol — propylene glycol methyl ether
098 Ethyl glycol — propylene glycol methyl ether
099 Thinner — acetone
100 Butanol/butyl acetate/toluene, etc. — propylene glycol ethers
101 Cyclohexanone/butyl acetate/aromatics — propylene glycol methyl ether
102 Isophorone/solvents — solvents
103 2-Butanone/toluene — 2-propanol/benzine
104 Hexane — heptane
105 Petroleum — "Nordkemi"
106 White spirit — C_{11} to C_{12} alkanes
107 Thinner — other liquid
108 Ethanol (pyridine) — ethanol (2-propanol)

DEGREASING AGENTS

109 Miscellaneous — miscellaneous
110 Trichloroethylene — water/solvents
111 Trichloroethylene — warm alkali
112 Trichloroethylene/lacquering by immersion — alkali/electrostatic powder lacquer
113 Trichloroethylene — alkali
114 Trichloroethylene — alkali
115 Trichloroethylene — alkali, 1,1,2-trichloro-1,2,2-trifluoroethane or trichloroethane
116 Trichloroethylene/trichloroethane — alkali
117 Organic solvents — alkali
118 Organic solvents — alkali
119 Hydrocarbons/high pressure — water/glycol/tap water pressure
120 Organic solvents — water-soluble solvents (enclosed)
121 Chloric solvents — alkali
122 Petroleum — alkali
123 Petroleum/cyclohexanol — warm water
124 Ethanol(den) — 2-propanol/water/ammonia
125 Methylene chloride — 1,1,1-trichloroethane
126 Trichloroethane — aliphatic mineral oils
127 Toluene/methanol, etc. (5-3) — acetone (4-1)
128 Ethanol (pyridine) — ethanol (2-propanol)
129 Solvent naphta 80/110 — odorless petroleum
130 Low-boiling organic solvents — high-boiling solvents or water-based
131 Organic solvents — *n*-paraffins

EXAMPLES

DISINFECTANTS
132 2-Propanol/nitrite/   Rodalon/water
    Rodalon/water

DETERGENTS
133 Acetone   ethanol

STAIN REMOVERS
134 1,1,1-Trichloroethane   1,1,2-trichloro-1,2,2-trifluoroethane

HIGH-PRESSURE CLEANING AGENTS
135 Solvents/high pressure   Alkali/tap water pressure

RUST INHIBITORS
136 Mineral oil/petroleum   aqueous

RUST REMOVERS
137 Solvents   aqueous agent

ABRASIVES
138 Quartziferous   burned kaolin
139 Flint   aluminum silicate

RELEASING AGENTS
140 Mineral oil/diesel oil, etc.   water
141 Mineral oil   water-based
142 Mineral oil/solvents   water/ester oil
143 Mineral oil   vegetable oil or water
144 Petroleum/train oil   ester oil
145 Trichloroethane/solvent naphta   trichlorofluoromethane/benzine

LUBRICANTS
146 Rinsing agent   water/surface active substances/benzoate or without lubricant

147 Silicone   soft soap
148 Petroleum   Exxol D 80

STABILIZERS
149 Phenylendiamine compound   butyl phenol compound

FILLERS
150 2-Component epoxy
151 Dry mortar
152 Toluene
153 Polysulfide joint filler

bitumen/cold molten metal
wet mortar
C_8 to C_{14} aromatics
silicone joint filler

MISCELLANEOUS
154 Ethanol (vegetable turpentine)
155 2-Propanol
156 Dibromoethane/petroleum
157 Crease-proof treatment
158 Shrink plastic/heat
159 Welding

ethanol (2-propanol)
borax/water
water
other crease-proof treatment
mechanical packaging
polyurethane joint filler

PESTICIDES AND HERBICIDES
160 Cerecit anti-moss
161 Phenoxyacetic acid herbicides

162 Chemical herbicides

authorized agent
fertilization/watering/cutting/
  weeding hoe
many kinds, e.g., plants covering
  the ground, flames

5.3 Examples

001   ASSESSMENT OF PRODUCTS USED IN A GARAGE

Professional group:	The automobile trade, motor-mechanics.
Working function:	General repairs of cars.
Cause:	A wish to remove the most hazardous products.
From product/process:	Shell coolant (ethylene glycol).
Solution:	Glad coolant, Mobil, or Brugsen (propylene glycol).
Method:	Toxicity assessment and vapor pressure.
Summary:	The garage used coolant of ethylene glycol. Replaced by propylene glycol coolant.
Satisfied:	Yes, because the new procedure is less hazardous and equally good.

EXAMPLES

002   COOLANTS

Professional group:	The automobile trade, motor mechanics.
Working function:	Filling and draining off coolant on buses.
Cause:	Knowledge of health hazard. Request from the Occupational Health Service and the safety board.
From product/process:	Ethylene glycol.
Solution:	Propylene glycol.
Method:	Toxicity assessment.
Summary:	The substitution was carried out in cooperation with the suppliers of the buses at the request of the Occupational Health Service and the safety board.
Satisfied:	Yes, because propylene glycol does not result in the chronic damages that ethylene glycol does. No, because the motor mechanics are encumbered with more irritant effects when they work with propylene glycol, because it is extremely hygroscopic and thus irritates the mucous membranes.

003   THE ADDITION OF SPIRIT TO COMPRESSOR UNITS LOWERING OF THE FREEZING POINT

Professional group:	Iron shipyards.
Working function:	The use of respiratory equipment fed by compressed air.
Cause:	Unpleasant smell of denaturant from the use of respiratory equipment fed by compressed air.
From product/process:	Spirit methylated by, respectively, pyridine and synthetic amyl alcohol.
Solution:	Spirit only methylated by synthetic amyl alcohol.
Method:	Assessment of smell.

Summary:	To counter formations of ice, in frosty weather, at valves, by use of pneumatic tools, spirit is added to the compressor unit. The compressor units also supply persons using respiratory equipment fed by compressed air. The irritating smell is first and foremost thought to be caused by the denaturant pyridine. In the future, spirit without pyridine is to be used.
Satisfied:	No, because some workers are still insecure about a possible harmful effect. It is recommended to use ethanol without denaturant.

004 COOLANTS/ANTIFREEZES

Professional group:	Engineering industry, unskilled.
Working function:	Filling coolant on petrol and diesel engines.
Cause:	Wish to find a product which does not stink but still has low toxicity.
From product/process:	Mobil Permazone PG, with propylene glycol.
Solution:	Coolant with propylene glycol: BP Thermo Liquid 56155 or BP Thermo Liquid S 56158.
Method:	Information from suppliers and TI.
Summary:	Replacement of coolant based on propylene glycol by an agent with a less unpleasant smell.
Satisfied:	No, because the products recommended also cause irritating smell, probably owing to the content of sodium benzoate, which has an unpleasant smell when oxidized.

005 CHEMICALS IN MORTAR—LOWERING OF THE FREEZING POINT

Professional group:	Bricklayers.

Working function:	Use of mortar.
Cause:	Irritant effects of pyridine in methylated spirit. Pyridine is used as a denaturant.
From product/process:	Spirit methylated by pyridine. The alternative methanol was out of the question because of toxicity.
Solution:	Spirit methylated by 2-propanol (isopropyl alcohol) as denaturant. Spirit is available with many different denaturants.
Method:	Assessment of smell.
Summary:	The denaturant of spirit, used for freezing-point lowering chemical in mortar, was changed from pyridine to 2-propanol. The smell of the product was thereby essentially improved.
Satisfied:	Yes, because the bricklayers liked the substitution.

006   VEHICLES IN HANGARS

Professional group:	The aircraft industry, unskilled and skilled.
Working function:	Transport.
Cause:	Irritating smell, noise, health hazard.
From product/process:	Vehicles with petrol and diesel engines.
Solution:	Vehicles, etc., driven by electricity.
Method:	—
Summary:	Many vehicles and self-propelling machines (e.g., ladders) are used in hangars. All with internal combustion engines, which entail irritating smell, noise, and health hazard. Can to some extent be replaced by machines/vehicles driven by electricity.
Satisfied:	Partly. It is a continuous process. Not all vehicles can be made electricity driven.

007 GRAFFITI CLEANING

Professional group: Cleaning, municipal semiskilled workers.

Working function: Cleaning of toilet doors, toilet walls, etc.

Cause: Risk of exposure to organic solvents.

From product/process: Miscellaneous (graffiti remover, among others) with a content of toluene, methylene chloride, methanol, etc.

Solution: Today scouring powder or TIV hand cleanser dependent on the properties of the materials used. The doors will perhaps be painted anew.

Method: —

Summary: To avoid contact with toxic organic solvents, a number of products were excluded, and common scouring powder or a special hand cleanser (with glycol content) was used instead.

Satisfied: Yes.

008 PAINT AND LACQUER REMOVER (SADOLIN HYDRO-SOLVER)

Professional group: Bricklayers and bricklayers' unskilled laborers.

Working function: Cleaning of lacquer on brick floors indoors.

Cause: Health hazard from solvents.

From product/process: Hydro-Solver, MAL code 5-3. Contains methylene chloride, ethanol/methanol. It is very fast.

Solution: Polyfilla Ludpasta is used, alkaline paint/ lacquer remover, containing calcium hydroxide, MAL-code 00-4.

Method: —

Summary: —

EXAMPLES   **85**

           Satisfied: Yes, because there are now less irritant effects both on the worker and the residents of the house.

009   COLOR REMOVER

    Professional group: Plastics industry—silk screen printing, unskilled.

    Working function: Washing of silk screen chases by the machine. (Wiping off.)

    Cause: The workers were afraid.

    From product/process: 87% acetone, 13% methyl ethyl ketone.

    Solution: 100% acetone.

    Method: —

    Summary: From a technical point of view, another solvent could be used, but the workers want something quick-drying (volatile), otherwise the piece rate and the scrap quantity will be affected. The problem of ventilation is being solved, but the admission of air is too small, which easily results in low pressure.

    Satisfied: Yes, because a pure product is better than using compounds. No, because there is no ventilation during the work.

010   STRIPPING OF AIRCRAFT WHEELS

    Professional group: Aircraft industry, unskilled.

    Working function: Stripping of aircraft wheels before inspection.

    Cause: Health hazard.

    From product/process: Immersion in an open container with stripper, containing among other things ethanolamine and phenol.

    Solution: Sandblasting with plastic granules in a closed cabin.

Method: Toxicity assessment.

Summary: Stripping in view of inspection of cracks, etc. Phenol caused irritating smell and health hazard.

Satisfied: No, because they forgot to ask the organizations before the purchase—a cabin with very bad ergonomic working conditions was acquired.

011   REMOVAL OF GRAFFITI FROM BUS SEATS

Professional group: Automobile trade, cleaning personnel.

Working function: Cleaning.

Cause: Considerations of a new product to remove graffiti.

From product/process: Hand cleanser.

Solution: The agent was not used.

Method: Toxicity assessment.

Summary: The company considered a switch over to a spray product with N-methyl-2-pyrrolidone.

Satisfied: Yes, because the product was not used.

012   REMOVAL OF PAINT FROM ENGINE PARTS

Professional group: Smiths.

Working function: Cleaning of engine parts in cleaning containers.

Cause: Irritating smell and uncertainty about health hazards.

From product/process: Ortho-dichlorobenzene with phenol activator.

Solution: Removal of paint in blast cabin with plastic granules as blasting agent.

Method: Toxicity assessment.

EXAMPLES 87

Summary: For removal of paint on engine parts *o*-dichlorobenzene is used with a phenol activator heated to 50 to 60°C; the paint may be removed by plastic sand in a blast cabin.

Satisfied: No, because it was not used for financial reasons.

013 AQUEOUS WOOD PRESERVATIVES

Professional group: Painting trade, painters.

Working function: Coating with aqueous wood preservative.

Cause: Smell, irritation, dizziness, headache.

From product/process: Wood preservative (a combination of acrylic/alkyd components) with approximately 2.5% volatile solvents (white spirit, 2-propanol and others). Code no. 0-1. The product was usually applied twice.

Solution: Two different wood preservatives (a combination of acrylic/alkyd components and acrylic component) with nonvolatile solvents

(propylene glycol and Texanol). Both coded 00-1.

Method: MAL code/toxicity assessment/threshold limit value/VHR.

Summary: A wood preservative with 2.5% volatile solvents was replaced by two different wood preservatives with nonvolatile solvents.

Satisfied: Yes, because the painters were satisfied.

014 PAINTING MATERIALS WITH ORGANIC SOLVENTS TO WATER-BASED PAINT

Professional group: Engineering shop, painters.

Working function: Maintenance of the interior of buildings.

Cause:	Dissatisfaction among the painters.
From product/process:	Painting of woodwork in buildings with paint MAL code 2-1.
Solution:	Replaced by paint with MAL code 0-1.
Method:	Literature, MAL code, assessment of smell.
Summary:	Paint with a high content of organic replaced by a water-based agent.
Satisfied:	Yes, because the painters are satisfied with the change.

015 AQUEOUS LACQUER

Professional group:	The painting trade, painters.
Working function:	Lacquering of parquet floors.
Cause:	Fatigue, headache, smarting eyes, dry throat, dizziness, and nausea.
From product/process:	Aqueous acrylic lacquer with toluene and two nonvolatile solvents. Code no. 0-1.
Solution:	Aqueous acrylic lacquer with only nonvolatile solvents. Code no. 00-1.
Method:	MAL code/threshold limit value/toxicity assessment/VHR.
Summary:	An aqueous acrylic lacquer with, among other things, toluene, has been proposed to be replaced by a product without volatile solvents.
Satisfied:	The substitution has not been carried out yet.

016 LACQUERING AND SOLVENT-BORNE PAINT PRODUCTS

Professional group:	The painting trade, painters.
Working function:	Application of lacquers and latex paints.

EXAMPLES 89

Cause:	Hazardous evaporation substances from lacquers and latex paints.
From product/process:	Lacquers and latex paints.
Solution:	Silicate paint.
Method:	—
Summary:	Study of alternative products for lacquers and latex paints. The conclusion: [to use] silicate paint, which does not evaporate hazardous substances to the extent lacquers and latex paints do.
Satisfied:	Yes, because the concentration of the solvents is essentially reduced during application as well as in the subsequent period in the room.

017   ASSESSMENT OF COLORS FOR SILK SCREEN PRINTING

Professional group:	Textile, silk screen printers.
Working function:	Printing of advertising signs on cardboard, paper, plastic.
Cause:	Wish to find a water-based product.
From product/process:	Common silk screen printing colors with solvents.
Solution:	Colors with a lower content of solvents (Aqua-jet from Pröll). The colors may be diluted with water.
Method:	Background knowledge.
Summary:	The task was initiated because of poor ventilation conditions. The workers wanted to have the colors with solvents removed, and it was technically possible.
Satisfied:	Yes, because the colors used now are less hazardous to health.

018 LACQUER FOR PLASTER MODELS

Professional group: Ceramics industry—porcelain, modelers.

Working function: Lacquering of plaster models—spraying.

Cause: The lacquer used contained lead chromate (carcinogenic).

From product/process: —

Solution: Development of lacquer without lead chromate by the supplier (Beckers), with organic pigment.

Method: List of carcinogenic substances.

Summary: Model lacquer containing lead chromate is replaced by a lacquer with corresponding properties and color, but without lead chromate.

Satisfied: Yes, because the lead chromate is removed, the lacquer is OK.

019 PAINTING OF CURBS

Professional group: Road painters.

Working function: Painting of yellow curb stripes.

Cause: Acute poisoning symptoms.

From product/process: Mercalin code no. 2-3, toluene, lead, chromate, solvent naphtha 80/110.

Solution: Change of process. Replacement of Mercalin 2-3 with colors without lead and chromate and fewer hydrocarbons.

Method: Toxicity assessment.

Summary: Acute poisoning symptoms as a result of painting yellow curb stripes. The process is changed from pneumatic sprayer to low pressure-airless with extension. The product is replaced and the symptoms disappeared.

Satisfied: Yes, because the irritant effects disappeared; control measurements satisfactory.

EXAMPLES   91

020   SPRAY PAINT

    Professional group: Automobile trade, painters.

    Working function: Spray paint.

    Cause: The Danish Working Environment Service.

    From product/process: Product coded 4-3.

    Solution: Product without lead and chromium—code 4-1 and improved ventilation in the cabin.

    Method: MAL code.

    Summary: Fraction sum before: dust: 110; solvent: 1.4. Fraction sum after: dust 0.77; solvent: 0.07.

    Satisfied: Yes, but would have liked a lower number before the hyphen.

021   SPRAY PAINT

    Professional group: The automobile trade, painters.

    Working function: Spray painting.

    Cause: The Danish Working Environment Service.

    From product/process: Product with lead and chromium.

    Solution: Product without lead and chromium.

    Method: MAL code.

    Summary: Fraction sum before: solvents: 0.89; dust: 67. Fraction sum after: solvent 0.39; dust: 0.87.

    Satisfied: Not informed.

022   CAR SPRAYING

    Professional group: Painters.

    Working function: Spray painting of car bodies.

    Cause: Reorganization of production brought on the possibility of change to other products.

From product/process: Acrylic lacquer, dissolved in thinner with ethyl glycol/methyl glycol and/or their acetates.

Solution: Acrylic lacquer dissolved in thinner without the glycol ethers/acetates mentioned above.

Method: Threshold limit value/VHR/toxicity assessment. MAL code cannot be used, as the system is based on threshold limit values from 1981.

Summary: Car lacquers dissolved in thinner which were otherwise technically comparable were assessed by their content of volatile glycol ethers with low threshold limit values. Products with these substances were excluded.

Satisfied: Yes, because it was a good solution from considerations of health. Also, technically the substitution seems successful.

023 UV-HARDENED LACQUER

Professional group: Electronics industry, unskilled.

Working function: Silk screen printing.

Cause: Assessment of possible health hazards by use.

From product/process: UV-lacquer with large evaporation of formaldehyde.

Solution: UV-lacquer with less or no evaporation of formaldehyde.

Method: Inquiry to the manufacturer when a degassing analysis showed more than 6 ppm formaldehyde at 20°C.

Summary: The manufacturer himself was not aware of the evaporation of formaldehyde, as it was not part of the recipe. The result of a meeting was that he asserted that he would find the source of the formaldehyde and limit it, or if possible,

	Satisfied:	eliminate it by a substitution in the product. He did not want his products to have a bad reputation.
		Yes, because this may have led to influence of the producer link.

024   ROAD MARKING

Professional group:	Municipalities, municipal workers.
Working function:	Paint for marking of asphalt and concrete.
Cause:	Problems of headache and nausea.
From product/process:	Aliphatic hydrocarbons: 5 to 20%; white spirit: 5 to 20%; toluene: 5 to 20%.
Solution:	Aliphatic hydrocarbons: >20%; propanol: 1 to 5%.
Method:	Threshold limit value.
Summary:	The solvent propanol is less hazardous to health than toluene.
Satisfied:	Yes, because the workers got a product less hazardous to health.

025   HANDWRITTEN SIGNS FOR SHOPS

Professional group:	Supermarket display artists.
Working function:	Use of ink for signs.
Cause:	Insecurity with the existing ink. Air pollution of 4 to 5 times threshold limit value 1985.
From product/process:	Ink for signs with ethyl glycol and ethanol.
Solution:	Two suppliers were requested to change the composition. This resulted (2 years later) in the development of ink with 50% ethanol and 50% water.
Method:	—

Summary: By request a supplier has changed ink for signs so that parallel measurements show: old ink: fraction sum 4.4 and 5.8. New ink: fraction sum: 0.14.

Satisfied: Yes, because the air pollution has been reduced to a minimum. The workers experience no irritant effects.

026 SHOWCARDS FOR SHOPS

Professional group: Supermarkets, display artists.

Working function: Showcards.

Cause: Irritant effects when cleaning letters.

From product/process: Organic solvent for cleaning of the letters.

Solution: Change to water-based ink from oil-based ink.

Method: —

Summary: Cleaning of showcards (letters) caused headache. Oil-based ink was used. A change to water-based ink makes it possible to clean showcards with water.

Satisfied: Yes, because cleaning solvents have been replaced by water. The workers have experienced no irritant effects since.

027 PRESERVATIVES IN PORCELAIN PAINTING

Professional group: The porcelain industry, porcelain painters.

Working function: Porcelain painting.

Cause: Suspicion of cause of hair loss.

From product/process: Borax.

Solution: No preservative.

Method: Toxicity assessment.

Summary:	In the department two painters lost tufts of hair. An assessment of the paints led to suspicion of borax.
Satisfied:	Yes, because the preservative could be avoided.

028 THE USE OF REACTIVE DYES IN TEXTILE DYE WORKS

Professional group:	Textile workers.
Working function:	Dyeing of textiles (cottons).
Cause:	Risk of allergic respiratory irritations from using reactive dyes in powder form.
From product/process:	Powder color.
Solution:	Aqueous dispersion of color.
Method:	—
Summary:	—
Satisfied:	Yes, because the risk of inhalation is eliminated.

029 DILUTION OF SILK SCREEN PRINTING COLOR WITH XYLENE

Professional group:	Ceramics industry.
Working function:	Spray painting of porcelain figures.
Cause:	Problems of headache and nausea.
From product/process:	The solvent xylene.
Solution:	Silk screen printing colors could as well be diluted with white spirit without change of the quality of the product.
Method:	Threshold limit value.
Summary:	—
Satisfied:	Yes.

030 SOLVENT FOR THE DILUTION OF LUBRICANT

Professional group:	Metal works, unskilled.
Working function:	Application of lubricant film to aluminium foil.
Cause:	Pronounced irritant effects and symptoms from the central nervous system and exceeded threshold limit values.
From product/process:	95% isopropyl alcohol + 5% lubricant.
Solution:	Ethanol denatured by 10% isopropyl alcohol instead of pure isopropyl alcohol.
Method:	Toxicity assessment.
Summary:	Pure isopropyl alcohol was replaced by ethanol, denatured by isopropyl alcohol, used for dilution of lubricant.
Satisfied:	Yes, because measurements showed results far below threshold limit values. Since then, a new technique has come into use. Dilution and the use of solvents is no longer necessary.

031 CHEMICALS FOR FLOCCULATION PLANTS

Professional group:	Slaughterhouses and meat industry, unskilled/smiths.
Working function:	Supervision and maintenance of the plant.
Cause:	Headache, slight illness.
From product/process:	Hydrocarbons are part of the flocculent.
Solution:	A flocculent without hydrocarbons.
Method:	—
Summary:	Replacement of a product containing a volatile/hazardous compound with a product which does not cause such vapors.
Satisfied:	Yes, because the new one has the same effect but does not entail the same risk.

EXAMPLES 97

032 HAND CLEANING

Professional group:	Engineering shop, engine fitters.
Working function:	Metalwork.
Cause:	Skin irritations of the workers.
From product/process:	Their hands were very dirty from handling metal goods. Hand cleanser with organic solvent is used for cleaning purposes.
Solution:	Use of working gloves, hand cream, and a mild soap.
Method:	Assessment of hand eczema and the dryness of the hands.
Summary:	Hand cleanser with organic solvent was replaced by improved hand hygiene, hand cream, and soap and hand cleanser with grains for special cases.
Satisfied:	Yes, because skin irritation was reduced.

033 HAND CLEANSER

Professional group:	Engineering shop, skilled and unskilled.
Working function:	Cleaning of hands after general engineering shop work.
Cause:	Skin problems, typically as exanthema.
From product/process:	PEVA-STAR containing coconut oil, citrus oils, lanolin derivatives, tensides, abrasives.
Solution:	SUPER PLUM.
Method:	Experiences of other workers and one's own.
Summary:	Because of skin problems with PEVA-STAR (owing to lanolin/citrus oils?) SUPER PLUM hand cleanser is chosen as replacement.
Satisfied:	Yes, because the new hand cleanser is sufficiently effective and the skin problems disappeared.

034   SEWING THREAD

Professional group: Clothing industry, seamstresses.

Working function: Sewing in impregnated waterproof cloth with impregnated thread.

Cause: 15 of 18 seamstresses had symptoms such as headache, vomiting, nausea, and diarrhea.

From product/process: Cloth and thread impregnated with evaporating formaldehyde, siloxane oil, and paraffin. No declaration, only verbal guarantee from an agent of the suppliers.

Solution: Cloth and thread replaced by other brands.

Method: Meetings lasting for hours with the participation of management, workers, suppliers, the Occupational Health Service, and the Danish Working Environment Service.

Summary: After 2–3 months the company replaced the materials causing the problems for the workers and production.

Satisfied: No, because I do not know what happened since, as the company got a new owner and developed a new policy of cooperation with the Occupational Health Service and later moved to the Occupational Health Service of Professional Groups.

035   INSULATION FIBER

Professional group: Ceramics industry, smiths.

Working function: Maintenance of machinery, kilns.

Cause: Discussion of asbestos, spring 1986.

From product/process: Asbestos fiber.

Solution: Insulation fiber without asbestos (Fiberfrax).

Method: Toxicity assessment.

EXAMPLES 99

Summary: —

Satisfied: Yes, because the workers were satisfied, and the asbestos was thrown out. No, because we have had no opportunity to make a thorough toxicity assessment of the new material and its possible secondary products at high temperatures.

036   BATH FOR TESTING RUBBER CONES FOR HIGH-VOLTAGE CABLE

Professional group: Unskilled.

Working function: Testing of the conductive properties of rubber cones by applying high voltage in a container with Freon 113 (nonconductive).

Cause: Irritant effects and measurements showing concentrations up to the threshold limit value.

From product/process: Freon 113.

Solution: Filling the container with oil.

Method: Toxicity assessment.

Summary: Oil may replace Freon 113 as a nonconductive liquid in a container for testing rubber cones for high-voltage cable.

Satisfied: Does not know—but the method has come into use.

037   PRESERVATIVES FOR PET FOOD

Professional group: Slaughterhouses and meat industry, unskilled.

Working function: Processing raw products for pet food.

Cause: Stomachache in periods when the preservative was used.

From product/process: Acetic acid.

Solution: Proposals of using, e.g., citric acid instead.

Method:	VHR
Summary:	—
Satisfied:	Does not know, as proposals for solution have not yet been tested in practice.

038 DRILLING AND CUTTING IN STAINLESS STEEL

Professional group:	Forge.
Working function:	Drilling and cutting in stainless steel.
Cause:	Fear of damages from the solvents of the drilling liquid.
From product/process:	Drilling liquid with almost 100% 1,1,1 tri-chloroethane.
Solution:	Drilling with rapeseed oil with a high content of erucic acid.
Method:	Toxicity assessment.
Summary:	Rapeseed oil functions well as coolant/lubricant. However, when stored for a longer time the rapeseed oil begins smelling bad. Rapeseed oil with a high degree of erucic acid is purchased from Århus oliemølle.
Satisfied:	Yes, because it functions, is considerably less toxic, and the workers are satisfied. No, because it is perishable.

039 COOLANTS/LUBRICANTS

Professional group:	The electronics industry, unskilled.
Working function:	Operator of cutting machine.
Cause:	Choice between two coolants/lubricants.
From product/process:	Coolant/lubricant based on mineral oil containing 32% unspecified di- and nitrilotri-ethanols, nitrite, and bactericide evaporating formaldehyde.

Solution:	Water-based coolant/lubricant with 15% mono- and 2% diethanolamine and bactericide, not splitting off formaldehyde.
Method:	Toxicity assessment, VHR, unsympathetic supplier, literature about conversions of coolants/lubricants.
Summary:	Di- and nitrilotriethanols may together with nitrite form nitrosamines. In the pH range approximately 6 to 11, formaldehyde may function as a catalyst in this process. Nitrosamines have proved to be carcinogenic in experimental animals. This among other things, and the fact that the other agent is based on mineral oil, made me prefer coolant/lubricant no. 2.
Satisfied:	Yes, because the manufacturer according to his own statement would not deal in carcinogenic substances and removed nitrite and nitrilotriethanol (in a fortnight). No, because screening of the machine is the only remedy to protect the operator from permanent damage.

040   COOLANTS/LUBRICANTS

Professional group:	Turners and smiths, among others.
Working function:	Turning, milling, etc.
Cause:	Skin irritations and smell caused by growth of microorganisms over the weekend.
From product/process:	Coolants/lubricants based on mineral oil.
Solution:	Synthetic and half-synthetic coolants/lubricants.
Method:	Toxicity assessments.
Summary:	Coolants/lubricants based on mineral oil have in several cases been replaced by whole and half-synthetic products.

Satisfied:	Yes, because skin irritation and smell, caused by growth of microorganisms over the weekend, have been removed. No, because the managers complain about shorter lasting tools.

041 COOLANT/LUBRICANT CONTAINING CHLORINATED PARAFFIN AS EP-ADDITIVE

Professional group:	Engineering shop, engine fitters.
Working function:	Grinding of big metal materials.
Cause:	The workers felt insecure about the substances in coolants.
From product/process:	Coolant with chlorinated paraffin was changed to coolant without chlorinated paraffin.
Solution:	Coolant without chlorinated paraffin, since the use of EP-additive was not a condition of the technical requirement.
Method:	Assessment of technical and occupational/ environmental requirements.
Summary:	Coolant with chlorinated paraffin was replaced by coolant without chlorinated paraffin, since there was no technical reason for chlorinated paraffin.
Satisfied:	Yes, because a substance was removed from the coolant. No, because there is still uncertainty about the other component substances.

042 COOLANT/LUBRICANT

Professional group:	Iron and metal industry. Unskilled, smiths.
Working function:	Operators of turning, milling, drilling, and grinding machines.
Cause:	Many developed eczema—considerable growth of bacteria.
From product/process:	With EP-additive and petroleum sulfonate (VB).

Solution:	Change to product with coconut fatty acid and boric acid complex.
Method:	Toxicity assessment, the experiences of other workers.
Summary:	Replacement of coolant/lubricant by a more ecological coolant/lubricant. At the same time a number of parameters are subject to continuous control.
Satisfied:	Yes, because the new one is better, both technically and from an ecological point of view.

043 COOLANT/LUBRICANT

Professional group:	—
Working function:	Turning/milling.
Cause:	Contains chlorinated paraffins as EP-additives.
From product/process:	Blasocut 2000 Universal.
Solution:	Houghton Hocut 2 M AL VBM.
Method:	Toxicity assessment, the experiences of others, problems with information from suppliers.
Summary:	Component substances of different coolants/lubricants assessed. The agent chosen functions well technically.
Satisfied:	Yes, because it is technically a good solution. VBM is especially foam reducing.

044 COOLANT/LUBRICANT

Professional group:	Skilled/unskilled.
Working function:	Miscellaneous metal working.
Cause:	Skin irritation.
From product/process:	CIMCOOL E5 STAR.

Solution:	FERMA U.
Method:	Technical assessment.
Summary:	Change from coolant/lubricant containing biocide to coolant/lubricant free from biocide.
Satisfied:	Yes, because skin irritation has disappeared.

045 COOLANT/LUBRICANT FOR GRINDING OF HIGH-SPEED STEEL (VB) AND CEMENTED CARBIDE TOOLS

Professional group:	Iron and metal/furniture, toolmaker.
Working function:	Grinding of tools from joiner's workshop.
Cause:	Workers with eczema in grinding workshop.
From product/process:	Coolant/lubricant probably dissolving cobalt.
Solution:	Coolant/lubricant which does not dissolve cobalt (Pers kemi: Meqqem-cob 8507).
Method:	Toxicity assessment.
Summary:	Later on it appeared that the cause of the eczema was not cobalt, but the substitution was kept.
Satisfied:	Yes, because the harmful effects on health might be reduced. No, because we have ourselves not (yet) had the opportunity to measure the cobalt strain before and after the substitution.

046 COOLANT/LUBRICANT

Professional group:	Plastics industry, smiths.
Working function:	Grinding, milling, turning, etc.
Cause:	Control of lubricant needed (concentration, pH, germ number).
From product/process:	Coolant/lubricant with nitrite (Cimcool E5 star).

EXAMPLES

Solution:	A coolant/lubricant without nitrite (Cimcool 5 star 40 or Cimplus D14).
Method:	Toxicity assessment (nitrite + amines = cancer).
Summary:	—
Satisfied:	Yes, because the risk of cancer is reduced. The smiths were satisfied. No, because a full toxicity assessment of the coolants/lubricants was not possible because of lack of information about the other content.

047 DRILLING-MILLING TASKS

Professional group:	Iron industry, engine fitters.
Working function:	Chipping processes working with SG iron.
Cause:	Exceeding of threshold limit value of phosphine.
From product/process:	Common coolant/lubricant.
Solution:	Synol with antiphosphine additive + exhaust.
Method:	Threshold limit value.
Summary:	The task has been solved in cooperation with Kemisk Værk Køge. Many other coolants/lubricants with phosphine additives have been tried, but none with such good results as Synol.
Satisfied:	Yes, because we now keep phosphines from 1/4 to 1/10 below the permitted threshold limit value.

048 SOLVENT FOR THE USE OF ANALYSIS

Professional group:	Chemical industry, laboratory assistants.
Working function:	Analysis of the intermediate *para*-nitrophenol.
Cause:	Carcinogenic.

From product/process:	10% benzene in isopropanol.
Solution:	10% toluene in isopropanol.
Method:	Toxicity assessment.
Summary:	A series of tests proved that the use of toluene instead of benzene did not influence the result of the analysis.
Satisfied:	Yes, because the fear of working with a carcinogenic substance was removed.

049 SOLVENTS FOR LABORATORY ANALYSIS

Professional group:	Chemical industry, laboratory assistants.
Working function:	Many different analyses.
Cause:	Carcinogenic.
From product/process:	Carbon tetrachloride and chloroform.
Solution:	Replaced by esters and ketones.
Method:	Toxicity assessment.
Summary:	As soon as it was recognized that the two substances were carcinogenic, they were removed from the laboratory and replaced by other substances.
Satisfied:	Yes, because the fear of working with carcinogenic substances was removed.

050 TLC RUNNING FLUID

Professional group:	Medicinal industry, laboratory assistants.
Working function:	Chemical analyses.
Cause:	A TLC running fluid—acetone 2, chloroform 18—has low threshold limit values.
From product/process:	—

EXAMPLES   107

Solution:	Tried to change to a TLC running fluid with a higher threshold limit value, ethanol 2, ethyl acetate 4, *n*-heptane 14.
Method:	Threshold limit values.
Summary:	—
Satisfied:	No, because the separating properties of the running fluid are not quite satisfactory.

051   TLC RUNNING FLUID

Professional group:	Medicinal industry, laboratory assistants.
Working function:	Chemical analyses.
Cause:	A TLC running fluid—chloroform 40, methanol 25, formic acid 7—has low threshold limit values.
From product/process:	(See cause.)
Solution:	Changed to a TLC running fluid, toluene 40, acetone 5, 100% acetic acid 4.
Method:	Threshold limit values.
Summary:	—
Satisfied:	Yes, because the running fluid has the same function but with SUBFAC figures, according to the SUBFAC calculation of the Occupational Health Institute, that are 100 times lower.

052   WEIGHING OF CARCINOGENIC POWDER

Professional group:	Chemical industry, laboratory assistants.
Working function:	Laboratory work.
Cause:	Determination of working process.
From product/process:	*o*-Phenylenediamine powder.

Solution: *o*-Phenylenediamine-dosed tablets.

Method: —

Summary: Laboratory assistants wanted to determine the working procedure of weighing carcinogenic powder. They began instead using dosed tablets which could be purchased (*o*-phenylenediamine).

Satisfied: Yes, because it does not dust and there is less refuse and cleaning.

053 PLASTIC GUTTERS

Professional group: Plumbing, plumbers.

Working function: Bonding of gutter elements.

Cause: Health hazard caused by the solvents in the adhesive.

From product/process: Content of tetrahydrofuran and methylene chloride code 3-1.

Solution: Some products are delivered with rubber strips. Mechanical joining without the use of adhesive.

Method: —

Summary: —

Satisfied: Yes, because the effect of solvents is entirely avoided.

054 TOE BOX SOFTENER

Professional group: Leather/footwear, unskilled.

Working function: Activation of glued-up toe boxes by means of toe box softener.

Cause: Headache, among other effects.

EXAMPLES

From product/process: The glue on the glued-up toe boxes was activated by toe box softener with 90% methylene chloride.

Solution: The glue on the glued-up toe boxes was replaced by a glue which could be activated by means of heat in ovens.

Method: Common sense.

Summary: —

Satisfied: Yes, because the workers were satisfied.

055 TANGIT—SUPPLIER OF ADHESIVES: SKANDINAVISK HENKEL A/S

Professional group: VVS (water/heat/sanitary installations) ventilation.

Working function: Bonding of hard PVC pipes. Large surfaces outdoors (ventilation piping in the earth).

Cause: Health hazards from solvents in the adhesive.

From product/process: Adhesive with 49% tetrahydrofuran, 20% cyclohexanone, and N-methyl-pyrrolidone and butanone.

Solution: Bonding is entirely avoided (mechanical joining + joint filler).

Method: —

Summary: Primarily owing to the economy of PVC pipe fittings, the company has used its creativity to develop pipe bends of galvanized sheets. These pipes are screwed mechanically together with straight pipes of PVC. The joint is sealed by means of joint filler.

Satisfied: Yes, because the solution was more economical and less hazardous to health.

056 JOINING OF WASTE PIPES FROM REFRIGERATED COUNTER

Professional group:	Plumbers.
Working function:	Plastic piping.
Cause:	Work with adhesive with solvents in closed rooms.
From product/process:	Tangit adhesive (methylene chloride) on PVC pipes. Cleaning with pure methylene chloride.
Solution:	Joining of ABS pipes by means of sleeves (no bonding).
Method:	Background knowledge + information from supplier.
Summary:	Plumbers bonded waste pipes for refrigerated counters with Tangit (methylene chloride) after cleaning with pure methylene chloride. We proposed pipes of ABS joined by sleeves.
Satisfied:	Yes, because the chemical product is entirely abolished.

057 CHANGE FROM ADHESIVES CONTAINING SOLVENTS TO WATER-BASED

Professional group:	Carpenter and joiner business, joiners.
Working function:	Use of adhesives.
Cause:	Wish to reduce the effects of solvents.
From product/process:	Adhesives based on solvents.
Solution:	Water-based adhesives.
Method:	MAL code, toxicity assessment, threshold limit values.
Summary:	Adhesives based on solvents replaced by water-based adhesives.
Satisfied:	Yes, because the users are satisfied.

EXAMPLES

058   ADHESIVE FOR FOAM PLASTIC MATERIAL

Professional group: Unskilled.

Working function: Bonding of foam plastic.

Cause: Working hygienic measurements/lack of exhaust.

From product/process: Adhesive containing among other things acetone, 1,1,1-trichloroethane, and toluene.

Solution: Water-based adhesive.

Method: The experiences of other workers, toxicity assessment.

Summary: —

Satisfied: Yes, because water-based adhesive satisfies the requirements of the finished product and improves working conditions.

059 ADHESIVE FROM LEATHER SCRAPS

Professional group: Leather manufacture, unskilled.

Working function: Bonding of leather/cardboard laminates as fixation before punching and sewing.

Cause: Evaporation of solvents, installation of ventilation.

From product/process: Contact adhesive with solvent naphtha 60/80, among other things, content of *n*-hexane.

Solution: Water-based contact adhesive.

Method: Comparative toxicity assessment.

Summary: Because of difficulties of getting satisfactory information from suppliers about the content, a technically suitable water-based adhesive was found by the Occupational Health Service, but the assessment was later made by the Register of Substances and Materials.

Satisfied: Yes, because the change to water-based contact adhesive was to a great extent a success.

060 FOAM PADDING ADHESIVES

Professional group:	Furniture industry.
Working function:	Bonding of train seats (spraying).
Cause:	Fear of vapors of solvents.
From product/process:	Adhesive based on organic solvent.
Solution:	Water-based adhesive + heat.
Method:	Threshold limit value + labeling.
Summary:	Spraying of foam padding adhesive based on organic solvent and in spraying box caused problems of vapor concentration in the air. Water-based adhesive and accelerated drying by heat eliminated this, but increased the effect of dust.
Satisfied:	Yes, because the effects of solvents were reduced. No, because instead an adhesive/dust problem may have arisen.

061 ADHESIVES

Professional group:	Bricklayers.
Working function:	Bonding of demo bricks to wooden sheets.
Cause:	Question: Is the product used, "Duracol 65," dangerous?
From product/process:	Water-based polyurethane adhesive containing MDI (diphenylmethane diisocyanate).
Solution:	Mounting adhesive based on solvents, satisfying technical requirements.
Method:	—
Summary:	Owing to technical requirements it is not possible to use water-based mounting adhesive, therefore a mounting adhesive based on solvents was chosen, MAL code 2–1.
Satisfied:	Yes, if sufficient exhaust is installed.

062 LAMINATION OF TEXTILES

Professional group:	Unskilled.
Working function:	Bonding of textiles.
Cause:	Organic solvents.
From product/process:	Roller application with adhesive containing: ethyl acetate, methyl proxitol, methyl proxitol acetate.
Solution:	Bonding with another method without organic solvents.
Method:	—
Summary:	(a) Change of process. (b) Change of materials.
Satisfied:	Yes, because the system works without internal and external environmental problems.

063 ADHESIVE FOR STICKING FOAM STRIPS TO TILES

Professional group:	Tileworks, unskilled.
Working function:	Application of adhesive and sticking of foam strips.
Cause:	Workers suffered brain damage.
From product/process:	Adhesive with solvent naphtha, ethyl acetate, and toluene. Diluent with acetone, solvent naphtha, isopropyl acetate, and toluene.
Solution:	Hot-melt adhesive consisting of 25% resins, 25% ethylene, 25% vinyl acetate, and 25% wax/paraffin.
Method:	Does not know.
Summary:	All solvents with the risk of damaging the brain were removed and a new adhesive on the basis of heat was used. The application of adhesive was automated and the smoke from the hot adhesive was removed by exhaust fans.
Satisfied:	Yes, because the working conditions were radically improved.

064 BONDING OF FOAM STRIPS TO TILES

Professional group: Tileworkers, unskilled.

Working function: Sticking of foam strips in automatically applied adhesive.

Cause: The adhesive used is too expensive.

From product/process: Hot-melt adhesive on the basis of resin with advanced opening: 25% resin, 25% vinyl acetate, 25% ethylene, 25% wax/paraffin.

Solution: Corresponding hot melt with retarded opening.

Method: Information from supplier.

Summary: Hot-melt adhesive with advanced opening replaced by corresponding less expensive adhesive with retarded opening. Owing to the retarded opening, the smoke from the hot adhesive could not be exhausted before it reached the workers.

Satisfied: No, because the workers are now exposed to vapors, causing irritant effects on the mucous membranes.

065 BONDING OF CARDBOARD BOXES

Professional group: Toy factory, unskilled.

Working function: Bonding of cardboard boxes.

Cause: Irritant effects on the workers.

From product/process: Bonding of cardboard boxes with hot-melt adhesive.

Solution: Use of self-adhesive tape.

Method: —

Summary: Cardboard boxes were closed with hot-melt adhesive, which was changed to self-adhesive tape and the irritant effects on the workers disappeared.

Satisfied: Yes, because the irritant effects disappeared.

EXAMPLES 115

066   ADHESIVE FOR ACRYLIC SHEETS

Professional group: Chemical industry, plastic, unskilled.

Working function: Bonding of acrylic advertising articles.

Cause: Order from the Danish Working Environment Service.

From product/process: Acrylic chips dissolved in chloroform.

Solution: Acrylics in ethyl acetate.

Method: —

Summary: The Danish Working Environment Service ordered that home-made acrylic adhesive must not be made using chloroform. Acrylic in methylene chloride was tried first. It caused more irritant effects on the workers than chloroform. Then ethyl acetate was proposed. We do not know the result.

Satisfied: We do not know.

067   WATER-BASED ADHESIVE FOR SHOEMAKING

Professional group: Custom-made shoemaking, shoemakers.

Working function: Bonding of soles on surgical shoes.

Cause: Wish to stop using adhesive based on solvents.

From product/process: Foss Fix 1222.

Solution: Casco contact adhesive 3885 V.

Method: —

Summary: The shoemaker has changed to Casco to avoid exhaust. We found out that the Casco adhesive contains approximately 3% toluene, in spite of deceptive information from the supplier. We still recommend exhaust.

Satisfied: No, because there are still solvents in the product, although it is better than Foss Fix 1222.

068   PVC ADHESIVES

Professional group: VVS (water/heat/sanitary installations), unskilled + some skilled.

Working function: Bonding of PVC pipes and fittings at the installation of water processing plants—typically plants which are not allowed to contain any metal.

Cause: Irritant effects from methylene chloride.

From product/process: Tangit bonding. Beforehand the surface is broken by wiping off with methylene chloride.

Solution: Pevicol adhesive based on N-methyl-2-pyrolidone. Boiling point 202°C, satisfactory toxicological data. The product requires less of the material. It was proposed that

the surfaces might be broken mechanically by means of emery cloth.

Method: Toxicity assessment, threshold limit values.

Summary: —

Satisfied: No, because the supplier of pipes would only guarantee durability if Tangit adhesive was used. He tried to dissuade me from my proposal in spite of satisfactory experiments at JTI and in the company. As many plants are installed abroad, the company did not dare to take the responsibility itself. But it wanted to try plants cautiously at home. The supplier of pipes also deals in Tangit adhesive. Pevicol is sold by the competitor.

069   BONDING OF HARD PVC

Professional group: Plumbers.

Working function: Bonding of pipes for compressors.

Cause: Irritant effects.

From product/process: Tangit (with methylene chloride) and pure methylene chloride for cleaning.

Solution:	Pevicol with *N*-methyl-pyrrolidone.
Method:	Toxicity assessment.
Summary:	Hard PVC pipes were bonded by Tangit (with methylene chloride) after cleaning with pure methylene chloride. Pevicol was tested. It satisfied the technical requirements of 1 atm. overpressure. After 2 months' work with Pevicol everybody is still happy; no irritant effects. Cleaning is performed with paper.
Satisfied:	Yes, because the irritant effects have gone.

070   PVC ADHESIVES

Professional group:	Medical industry, metal.
Working function:	Occasional bonding of PVC tubes for aggressive liquids.
Cause:	Irritant effects from the solvents in the adhesive.
From product/process:	—
Solution:	Pevicol adhesive.
Method:	Toxicity assessment, threshold limit values.
Summary:	—
Satisfied:	Yes, because the smiths were satisfied—both with the technical function and that the irritating smell disappeared.

071   BONDING OF PLASTIC

Professional group:	Shipyard—plumbers, model engineers.
Working function:	Bonding of plastic models and joining of plastic gutters.
Cause:	Use of hazardous substances.
From product/process:	Methylene chloride and tetrahydrofuran.
Solution:	*N*-methyl-pyrrolidone.

Method: —

Summary: —

Satisfied: Yes, because the substance is less harmful.

072 PVC ADHESIVE

Professional group: PVC factory.

Working function: Bonding of PVC pipes (drain pipes, cesspools, transition pieces, among others).

Cause: —

From product/process: Adhesive with tetrahydrofuran, cyclohexanone, methyl ethyl ketone.

Solution: Adhesive with N-methyl-pyrrolidone.

Method: VHR + toxicity assessment.

Summary: The company has taken steps to replace the adhesive with mechanical joining of the materials, as the new adhesive on the first day had already caused tingling in arms and fingers (because of skin contact).

Satisfied: At first yes, because everything seemed to go smoothly and be successful. Later no, because signs of absorption and poisoning appeared.

073 PVC ADHESIVE

Professional group: Plastics industry, unskilled.

Working function: Bonding of PVC plastic components.

Cause: Illness caused by adhesive + cleaning liquid.

From product/process: PVC adhesive based on tetrahydrofuran, cyclohexanone and cleaning liquid (methylene chloride).

Solution: PVC adhesive based on N-methyl-2-pyrrolidone (NMP), no cleaning liquid.

Method:	MAL code (factor 3 difference), VHR (factor approximately 200 difference). Some working hygienic measurements (performed by others) show very low fraction sums when working with this adhesive.
Summary:	—
Satisfied:	Yes, because much lower fraction sums may be obtained with working hygienic measurements. No, because toxicity assessment of NMP (e.g., teratogenic effect dependent on dose in mice/rats, high skin permeability) may make this substitution unwarranted.

074 BONDING OF PVC

Professional group:	Manufacture of plastic articles, skilled and unskilled.
Working function:	Bonding of PVC windows.
Cause:	Vapors hazardous to health from adhesive with solvents.
From product/process:	PVC adhesive for pipes 1268 (Sadofoss A/S): >20% tetrahydrofuran, 1 to 5% cyclohexanone, polyvinyl chloride copolymer.
Solution:	Pevicol (Rønne Mikkelsen A/S). *N*-methyl-2-pyrrolidone.
Method:	Toxicity assessment.
Summary:	Adhesive with solvents which are hazardous to health replaced by adhesive containing less substance hazardous to health (*N*-methyl-2-pyrrolidone). The adhesive is used for PVC materials which are to have great strength and be weatherproof.
Satisfied:	Yes, because the workers feel a considerable improvement of their working environment in spite of lack of establishment of mechanical ventilation.

075 PVC BONDING

Professional group: Plastics.

Working function: Bonding of PVC plastics.

Cause: Reorganization of the working process.

From product/process: PVC adhesive with methyl ethyl ketone.

Solution: PVC adhesive with N-methyl-2-pyrrolidone.

Method: Toxicity assessment.

Summary: In connection with a total assessment of the bonding process the adhesive was replaced by adhesive with another solvent.

Satisfied: Yes.

076 PLASTIC ADHESIVES

Professional group: Toy factory, unskilled.

Working function: Model making.

Cause: Damaged health of workers.

From product/process: Adhesive based on butanone.

Solution: Adhesive based on N-methyl-2-pyrrolidone.

Method: VHR.

Summary: Adhesive based on butanone was substituted by a product based on N-methyl-2-pyrrolidone on the basis of an assessment of the evaporation of, respectively, butanone and N-methyl-2-pyrrolidone of the adhesive.

Satisfied: No, because the new one does not function technically. A more detailed assessment of the toxicology of N-methyl-2-pyrrolidone (skin absorption, irritation, teratogenic properties) also had negative results.

077 LEATHER ADHESIVES

Professional group:	Leather industry, unskilled.
Working function:	Bonding of leather coats.
Cause:	Fear of health hazards from solvents in the adhesive.
From product/process:	Adhesive with 40% heptane, 20% toluene, and 18% ethylene acetate.
Solution:	Adhesive with 82% heptane, 4% toluene, and 1% ethanol.
Method:	VHR
Summary:	The evaporation of solvents from two different types of adhesive with toluene and heptane is assessed on the basis of VHR. The adhesive with the least toluene was chosen.
Satisfied:	Yes, because the workers are fully satisfied.

078 COUNSELING ABOUT EPOXY ADHESIVE IN A NEW PRODUCTION PROCESS

Professional group:	Textile industry, unskilled.
Working function:	Manufacture of reeds.
Cause:	Introduction of new method for financial reasons.
From product/process:	Reeds bonded with epoxy.
Solution:	Method of production indicated.
Method:	Toxicity assessment, evaporation.
Summary:	Introduction of a new product with epoxy adhesive was attempted, but the workers protested. We tried to introduce an adhesive/lacquer which is less hazardous to health but the product did not satisfy the technical requirements. Method of production was thereafter abandoned.

Satisfied: Yes, because a process of production which could be hazardous to health was not introduced. No, because we did not find a method which could be used.

079 IMPROVED ADHESIVE PROPERTIES OF ASPHALT

Professional group: Unskilled.

Working function: Repair of road asphalt (laying out of asphalt).

Cause: Irritating smell causing slight illness when laying out asphalt.

From product/process: Asphalt with improved adhesive properties.

Solution: Removal of improving agent from the asphalt.

Method: Toxicity assessment and experiences of others.

Summary: In hot sultry weather the work might cause nausea and indisposition. The improving agent was removed and the irritant effects have not reappeared.

Satisfied: Yes, because the illness has disappeared.

080 ELECTRONIC SOLDERING

Professional group: Toy factory, designer.

Working function: Electronic soldering.

Cause: Wish to avoid lead and colophony, and furthermore, washing off flux residues by organic solvents. Product safety.

From product/process: —

Solution: Automatic soldering with silver solder. Water-washable flux.

Method: Toxicity assessment, experiences of others.

EXAMPLES 123

Summary: With the establishment of a new job, the possibilities of avoiding known harmful effects of electronic soldering were examined. A silver solder without lead was chosen, together with water-washable flux. The soldering procedure is automatic, with exhaust from a totally enclosed system.

Satisfied: Yes, because it is technically successful. No reported irritant effects.

081   CLEANING AGENT FOR INJECTION-MOLDING MACHINES

Professional group: Toy factory, unskilled.

Working function: Cleaning of worm in injection-molding machine.

Cause: Irritant effects on workers because of evaporation.

From product/process: Cleaning agent based on polymethacrylate.

Solution: Cleaning agent consisting of 45% abrasive + 30% mineral oil + 17.6% water + 7% tensides.

Method: Toxicity assessment, experiences of other workers, adviser's assistance.

Summary: Cleaning agent based on polymethacrylate was replaced by a product based on an inorganic abrasive and mineral oil. The new product was technically not satisfactory.

Satisfied: No, because the product did not function technically.

082   CLEANING OF MACHINES

Professional group: Sweets, unskilled.

Working function: Machine operator.

Cause: Order from the Danish Working Environment Service about removal of chlorinated solvents from the cleaning process.

From product/process:	"Drycleaner 77" with trichloroethane.
Solution:	Several alkaline water-based products were tested before "A-Z renser 12" was accepted as a product of substitution. (Water, alkylarenesulfonate ethoxylate, potassium tripolyphosphate, max. 5% KOH, alkane sulfonate, silicates, isopropyl alcohol, phosphate, glycol).
Method:	Toxicity assessment.
Summary:	It is difficult to remove combinations of gum base, sugar dust, and talc powder from the machines in production/packing room without leaving a greasy film if water-based solvents are used. "A-Z renser 12" succeeded, but with the limitation that the new substance is locally irritant. Further substitutions may be possible.
Satisfied:	Yes, because the machine could be cleaned without using trichloroethane, although it takes a longer time with the new product.

083 CLEANING OF SPINNING MACHINES

Professional group:	Textiles, unskilled.
Working function:	Cleaning of machines.
Cause:	Risk of damages from using organic solvents.
From product/process:	1,1,2-trichloro-1,2,2-trifluoroethane.
Solution:	Change to alkaline degreasing.
Method:	—
Summary:	—
Satisfied:	Yes, because organic solvents are removed. Gloves afford complete protection.

EXAMPLES   125

084   CLEANING

 Professional group: Telecommunications, technicians.

 Working function: Cleaning of telex equipment.

 Cause: Wish to reduce the use of Freon.

 From product/process: Arklone P (1,1,2-trichloro-1,2,2-trifluoroethane).

 Solution: "Fective" brand detergent—water-based, alkaline (NaOH) detergent.

 Method: Toxicity assessment + good experiences with Fective for other cleaning purposes.

 Summary: The attempt at replacing a Freon detergent with an ecological water-based detergent was unsuccessful because the technical requirements were not satisfied.

 Satisfied: No, because Fective could not satisfy the technically specified requirements.

085   CLEANING OF FIRE ALARM (ION DETECTOR)

 Professional group: Electricians.

 Working function: Ion detector (fire alarm) cleaned to prevent false alarms.

 Cause: Headache, fatigue, and slight illness at the end of the day's work.

 From product/process: Freon TF = Freon 113 = 1,1,2-trichloro-1,2,2-trifluoroethane.

 Solution: Ajax for Windows.

 Method: Toxicity assessment, VHR.

 Summary: Instead of cleaning the alarm with Freon TF it is possible to use Ajax for Windows. However, fire insurance companies make trouble, so it is not possible to change straight away.

Satisfied: Yes, because the irritant effect was reduced. No, because certain suppliers do not accept the substitution with reference to insurance responsibility.

086 SPRINKLING LIQUID

Professional group: Transport, engine drivers.

Working function: Engine driver.

Cause: Reservoir placed in the engine driver's cab, spillage causes, among other things, considerable evaporation of 2-propanol.

From product/process: 2-Propanol in water.

Solution: 1,2-Propandiol (propylene glycol) in water.

Method: Toxicity assessment.

Summary: The solvent in sprinkling liquid was changed from 2-propanol to 1,2-propandiol. Then the irritating smell in the engine driver's cab disappeared.

Satisfied: Yes, because the engine drivers liked the substitution. No, because there are slight foam problems.

087 TECHNICAL CLEANING OF AIRCRAFT

Professional group: Aircraft industry, unskilled.

Working function: General cleaning of aircraft externally and cleaning of tectyl before inspection.

Cause: Use of mask. Neurotoxicity. Skin irritations.

From product/process: C-11 to C-12 *n*-alkane compound.

Solution: Aqueous, slightly alkaline product.

Method: Toxicity assessment.

EXAMPLES 127

Summary:	Use of mask necessary because of neurotoxicity while working in wheel wells.
Satisfied:	Yes, partly, but it is at present not proved if the product contains, e.g., 2-ethoxyethanol. Mask is unnecessary.

088 CLEANING OF CARPET LOOMS

Professional group:	Textiles, engine fitters + mechanics + engineer.
Working function:	Cleaning of looms and workshop.
Cause:	Headache after cleaning.
From product/process:	White spirit.
Solution:	Alkaline degreasing.
Method:	Change from organic solvents to alkaline degreasing.
Summary:	—
Satisfied:	Yes, because the organic solvents are removed. Gloves now protect fully.

089 PETROLEUM

Professional group:	Plastics industry, engine fitter.
Working function:	Cleaning of machines.
Cause:	Common replacement of organic solvents.
From product/process:	Petroleum.
Solution:	Alkaline industrial cleaning agent, conc. 2% (pH = 10.2) in solution to be used.
Method:	Assessment of health hazard.
Summary:	Petroleum has been replaced by alkaline industrial cleaning agent used in a solution of 2% (pH = 10.2). Dilution with warm water for cleaning purposes and application with a brush. Gloves are worn.

Satisfied: Yes, because professional group and safety representatives are satisfied.

090 CLEANING OF MOLDS

Professional group: Accumulator industry, unskilled.

Working function: Casting of lead gratings.

Cause: Corrosion hazard.

From product/process: Cleaning with NaOH solution.

Solution: Washing with warm soapy water.

Method: Experiences of other workers.

Summary: When molds are cleaned, the NaOH solution is replaced by warm soapy water. It is used in other companies with satisfactory results, also as concerns the maintenance of the molds.

Satisfied: Yes, because the hazard is eliminated, and the workers are satisfied.

091 CLEANING OF CONVEYOR BELTS

Professional group: Rubber factories.

Working function: PVC conveyor belts are cleaned manually (with a cloth) before PVC stops are welded on (ultrasound).

Cause: Fear of damages from solvent.

From product/process: Spirit (methylated).

Solution: Water-based cleaning agent, neutral pH. Others' and own experiences.

Method: —

Summary: The method was tried, and test weldings proved not to be durable. Therefore additional cleaning by water was necessary.

Satisfied: Yes, but the workers and the management thought it was a bad idea, because the new agent evaporated more slowly and it was necessary to polish afterwards with a cloth and clean water to avoid a film of cleaning agents. It took too much time.

092 CLEANING LIQUID

Professional group: Glass work, unskilled.

Working function: Cleaning of tools which have been used for insulation (molten glass).

Cause: Fear of health hazards.

From product/process: Cleaning with methylene chloride.

Solution: Toluene.

Method: Toxicity assessment.

Summary: Methylene chloride is replaced by toluene.

Satisfied: Yes, because at the same time ventilation has been established and gloves are used during the work and the workers are instructed in detail about the hazards of toluene.

093 CLEANING OF ACRYLATE PLASTICS

Professional group: Unskilled.

Working function: Acrylate coating is removed by methylene chloride before control.

Cause: Irritant effects on the workers, headache, skin irritation.

From product/process: Immersion in methylene chloride (followed by scraping off of the swelled plastic layer).

Solution:	Replacement of methylene chloride by *N*-methyl-2-pyrrolidone, ethyl acetate, or butyl acetate.
Method:	Parameters of solubility, VHR.
Summary:	On the basis of parameters of solubility, other solvents were proposed with almost the same property of dissolving acrylate plastics and with considerably less health hazard.
Satisfied:	Not concluded.

094 CLEANING CHEMICALS FOR OCTANE RATING ANALYSES

Professional group:	Oil trade, refinery technicians.
Working function:	Cleaning liquid for octane rating analyses.
Cause:	Fear of chloroform.
From product/process:	Chloroform.
Solution:	Ethanol.
Method:	—
Summary:	As concerns health, it is considered more warrantable to use ethanol rather than chloroform.
Satisfied:	Yes, because ethanol is less hazardous to health than chloroform.

095 CLEANING OF NOZZLES OF ACRYLIC PLASTIC RESIDUES

Professional group:	Unskilled.
Working function:	Nozzles are cleaned in ultrasonic bath.
Cause:	New type of nozzles requires more effective cleaning agent. Opposition from the safety board to the use of chlorinated solvents.
From product/process:	Ultrasonic bath with 1,1,1-trichloroethane.

Solution:	Ultrasonic bath with water-based, basic solution (it may be warm) or *N*-methyl-2-pyrrolidone.
Method:	VHR, parameters of solubility.
Summary:	Old type of nozzle cleaned in isopropyl alcohol. This may not be used in ultrasonic bath because of the danger of fire.
Satisfied:	Not concluded.

096 CLEANING LIQUID FOR PU BONDING WORK

Professional group:	Stone-working industry, unskilled.
Working function:	Cleaning of surplus adhesive.
Cause:	—
From product/process:	From product with 1,1,1-trichloroethane.
Solution:	Product instead containing glycol ether, which has also lower vapor pressure.
Method:	Toxicity assessment.
Summary:	Cleaning liquid with 1,1,1-trichloroethane replaced by a cleaning liquid which is less hazardous to health.
Satisfied:	Yes, because the product is less hazardous to health.

097 WASHING AGENT

Professional group:	Plastics industry, typographers, and unskilled.
Working function:	Typography. Washing.
Cause:	Problems of complying with the threshold limit value.
From product/process:	Ethyl glycol.
Solution:	Propylene glycol monomethyl ether.

Method: Various literature sources on toxicity.

Summary: —

Satisfied: Yes, because workers have expressed their satisfaction.

098 WASHING OF AIRCRAFT

Professional group: Aircraft industry, unskilled.

Working function: Cleaning of aircraft on the outside.

Cause: Risk of reproductive toxicity.

From product/process: Soap product with 2-ethoxyethanol.

Solution: Corresponding product with dipropylene glycol methyl ether.

Method: Toxicity assessment.

Summary: Letter to the manufacturer with a description of the problem and proposal of chemical for substitution (e.g., propylene glycol ethyl ether). Approximately 1 month later the product was changed.

Satisfied: Yes, because reproductive toxicity risk has been eliminated; moreover, the neurotoxicity risk has been reduced approximately 50 to 100 times.

099 CLEANING

Professional group: Porcelain, plasterers.

Working function: Cleaning of plaster molds.

Cause: Measurements/fear of damages.

From product/process: Thinner.

Solution: Acetone.

Method: Toxicity assessment.

EXAMPLES   133

Summary:	—
Satisfied:	Yes.

100   CLEANING LIQUIDS

Professional group:	Toy factory, unskilled.
Working function:	Cleaning of equipment.
Cause:	Irritant effects on the workers.
From product/process:	Cleaning liquid with, among other things, butanol, butyl acetate, 2-propanol, toluene, and xylene.
Solution:	A product based on propylene glycol compounds.
Method:	Vapor hazard index, toxicity assessment.
Summary:	Two cleaning liquids were assessed as to evaporation and toxicology. A product based on propylene glycol compounds was chosen because of its low evaporation and more "attractive" toxicology.
Satisfied:	Yes, because the irritant effects on the workers were reduced. Measurements showed considerably lower fraction sums.

101   CLEANING LIQUID

Professional group:	Plastics industry, unskilled.
Working function:	Washing of dabbing machine.
Cause:	Headache/slight illness. Fear of health hazard from former cleaning liquid.
From product/process:	Cleaning liquid with cyclohexanone, butyl acetate, C_9 to C_{14} aromatics.
Solution:	Cleaning liquid of propylene glycol methyl ether.

Method:	Toxicity assessment. Basic book of occupational medicine, part II.
Summary:	Evaporation from former cleaning liquid was unbearable. Cleaning liquid of propylene glycol methyl ether was found, estimated from basic book of occupational medicine, part II. At the same time the exhaust was improved, and the importance of avoiding skin contact was announced.
Satisfied:	Yes, because the operators like the product better. No, because glycol ethers are not quite harmless: the operators must be trusted not to forget the permeability of skin.

102 CLEANING DILUENT

Professional group:	Printers—tin packaging, unskilled.
Working function:	Washing off of lacquer machines.
Cause:	The supplier would not mix it for us. We will have to do it ourselves in the future.
From product/process:	66% Shellsol AB + 30% MIBK + 4% isophorone.
Solution:	66% Shellsol AB + 34% MIBK.
Method:	Our own laboratory estimated that isophorone was unnecessary.
Summary:	After pressure from outside we succeeded in getting a simplified mixture of solvents and perhaps the danger of combination effects is reduced.
Satisfied:	Yes, because the labeling was changed from Xi to F.

103 CLEANING OF COLOR TRAYS

Professional group:	Print on plastic film.

Working function:	Cleaning of the color trays of the printing machine.
Cause:	Fear of damages from the solvents in the cleaning liquid.
From product/process:	Cleaning liquid with 1 part methyl ethyl ketone and 1 part toluene.
Solution:	Cleaning liquid with 3 parts isopropanol and 1 part solvent naphtha 80/110.
Method:	Toxicity assessment.
Summary:	—
Satisfied:	Yes, because the workers are satisfied with the less unpleasant and well working cleaning liquid.

104   n-HEXANE FOR WASHING PURPOSES

Professional group:	Tape production, unskilled.
Working function:	Washing of roller with organic solvent, among other situations in case of suspension of operations.
Cause:	Counseling about instructions. The Occupational Health Service assessed processing chemicals: risk of synergism between methyl ethyl ketone and n-hexane. Measurement showed considerable exposure.
From product/process:	Hexane, distillation interval 65 to 69°C, contains a good deal of n-hexane.
Solution:	Heptane compound of hydrocarbons, distillation interval 94 to 99°C.
Method:	Boiling point interval + toxicity assessment.
Summary:	Tape is produced by applying a plastic film to a layer of adhesive. In case of suspension of operations the rollers must be washed with an

organic solvent to prevent them from getting sticky. Formerly an agent with *n*-hexane was used, while methyl ethyl ketone was used nearby in a printing process. Owing to strict technical requirements it was only possible to change to a heptane compound.

Satisfied: Yes, because things had improved. No, because exposure during work was still high. The company would not accept exhaust, but ordered masks which were not used. The company has now closed.

105   CLEANING LIQUID

Professional group: Accumulator company, unskilled.

Working function: Cleaning of objects polluted by lead oxide.

Cause: A wish to get rid of a substance which is hazardous to health.

From product/process: Petroleum was used for cleaning purposes with subsequent irritant effects and possible harmful effects.

Solution: Acetic acid was proposed, but the company found another product: Nordkemi 413. The effect of acetic acid was not tried.

Method: Toxicity assessment.

Summary: Petroleum was used for cleaning of lead oxide on surfaces. Petroleum was replaced by a less toxic product: Nordkemi 413.

Satisfied: Yes, because the workers are satisfied with the solution.

106   TECHNICAL CLEANING OF AIRCRAFT

Professional group: Aircraft industry, unskilled.

Working function: General cleaning of aircraft on the outside and cleaning of tectyl before inspection.

Cause:	Use of mask. Neurotoxicity. Skin irritation. Smell irritating to other workers.
From product/process:	White spirit with 17% aromatics.
Solution:	Solvent of C_{11} to C_{12} n-alkane-compound.
Method:	Toxicity assessment + VHR.
Summary:	Aircraft workers must wear mask + special working clothes. Others working on the aircraft at the same time are much irritated by the smell.
Satisfied:	Yes, because there are now fewer smell irritations in the surroundings, and lower concentrations because of slower evaporation. No, because it is still necessary to wear a mask.

107   CELLOSOLVE

Professional group:	Silk screen printers, unskilled.
Working function:	Printing on materials.
Cause:	Reduce the effect of solvent.
From product/process:	Thinner.
Solution:	Use of another kind of cleaning liquid.
Method:	Toxicity assessment, labeling.
Summary:	Thinner was replaced by another cleaning liquid. Measurements showed low concentrations of solvent.
Satisfied:	Yes, because the effect of the solvent was drastically reduced.

108   CLEANING LIQUID

Professional group:	Ceramics industry, molds.

Working function:	Production of molds.
Cause:	Odor and toxicity assessment of spirit denatured by pyridine.
From product/process:	Spirit denatured by pyridine.
Solution:	Spirit denatured by 2-propanol (isopropyl alcohol).
Method:	Odor, toxicity assessment.
Summary:	—
Satisfied:	Yes, because the odor improved, and the company was satisfied.

109 DEGREASING OF ENGINE PARTS

Professional group:	Garages, motor mechanics.
Working function:	Degreasing or cleaning of engine parts or whole motors.
Cause:	Question of health hazard in garages.
From product/process:	Products based on organic solvents.
Solution:	List of choice of products of approximately 30 items, listed according to health hazard with the least hazardous first. For a given task the first ones in the list are tried first. If the result is not good enough, a product further down the list is chosen.
Method:	Toxicity assessment.
Summary:	As an examination project in DIA-K, Karin Larsen made a substitution list of the cleaning agents of the automobile trade. The project contains a method of substitution, lists of choices, and data of products and component substances.
Satisfied:	Yes.

EXAMPLES

110   CLEANING

Professional group:	Iron and metal, engine fitters.
Working function:	Cleaning of oiled metal.
Cause:	Damages from trichloroethylene (tri).
From product/process:	Trichloroethylene plant.
Solution:	Reorganization to enclosed rotating flushing plant with water-based solvent.
Method:	—
Summary:	Reorganization from enclosed tri plant to water-based plant.
Satisfied:	Yes, because trichloroethylene is removed.

111   DEGREASING OF METAL

Professional group:	Iron and metal, unskilled.
Working function:	Degreasing of metal materials for intermediate stores.
Cause:	Acute poisoning.
From product/process:	Tri* degreasing.
Solution:	Basic degreasing and rust inhibition.
Method:	Toxicity assessments.
Summary:	—
Satisfied:	Yes, because the workers again feel secure with the working procedure.

112   FINISHING TREATMENT OF BICYCLE FRAMES

Professional group:	Metal, painters.
Working function:	Tri degreasing and dip coating.

---

* Tri = trichloroethylene

Cause: Accident of acute poisoning.

From product/process: Tri degreasing and dip containers with paint.

Solution: Centrifugal cleaning (1), basic degreasing (2), zinc phosphatizing (3), electrostatic sprayable powder coating (4); (2) and (3) in closed system.

Method: Toxicity assessments, VHI, etc.

Summary: Organic solvents have been replaced by less hazardous chemicals and working methods.

Satisfied: Yes, because the workers are satisfied, but the solution has been expensive.

113 DEGREASING IN THE METAL INDUSTRY

Professional group: Agricultural machine factory, smiths, and unskilled.

Working function: Degreasing.

Cause: Original trichloroethylene degreasing caused irritant effects.

From product/process: Trichloroethylene.

Solution: Alkaline degreasing.

Method: Assessment + measurement.

Summary: Trichlor degreasing containers rebuilt for alkaline degreasing.

Satisfied: Yes, because the risk of inhalation of trichlor and the increased risk in case of fire or failure of the ventilation has been removed. The painting department states that the new degreasing method is better than the former one.

114 FROM TRICHLOROETHANE DEGREASING TO WATER-BASED ALKALINE DEGREASING

Professional group: Iron and metal, unskilled.

EXAMPLES   141

Working function: Degreasing of materials.

Cause: Elimination of the use of trichloroethane.

From product/process: Degreasing with trichloroethane.

Solution: Use of alkaline detergents. Tailor-made solutions as to plants/chemicals in close cooperation with suppliers.

Method: All.

Summary: Replacement of trichloroethane with alkaline detergents for degreasing.

Satisfied: Yes, because the harmful effect from organic solvents is eliminated.

115   DEGREASING AGENTS

Professional group: Chemical industry, unskilled.

Working function: Manual degreasing of conveyor belts before bonding.

Cause: Study/measurement of effects.

From product/process: Trichloroethylene on cloth.

Solution: Three possibilities used according to the type of pollution: (1) water-based (sodium silicate); (2) Freon 113; (3) 1,1,1-trichloroethane.

Method: Threshold limit value, measurements of concentration in the air.

Summary: Measurements during degreasing with various solvents showed, as compared with trichloroethylene, a reduction of, respectively, 2.4 times by use of 1,1,1-trichloroethane and 8.4 times with Freon 113.

Satisfied: Yes, because the effects were essentially reduced.

116 DEGREASING OF MATERIALS BEFORE LACQUERING

Professional group:	Painter, unskilled.
Working function:	Cold degreasing of metal materials by immersion in containers before lacquering. Drying in the room.
Cause:	The workers got stomachaches because of the degreasing agent.
From product/process:	Degreasing agent for the most part consisting of trichloroethylene, diluted with a product for the most part consisting of 1,1,1-trichloroethane.
Solution:	Alkaline degreasing with a brush and rinsing with water. The agent contains, among other things, sodium metasilicate.
Method:	Common sense.
Summary:	The workers seldom used cold degreasing as they fell ill. Instead they used alkaline degreasing. They entirely abandoned the first method.
Satisfied:	Yes.

117 DEGREASING IN THE GRAPHIC INDUSTRY (TEXTILE TRANSFER)

Professional group:	Textile transfer printing, unskilled.
Working function:	Cleaning of rollers (stencils).
Cause:	Irritant effects by the use of solvents.
From product/process:	Hand cleaning with flammable organic solvents.
Solution:	Alkaline degreasing in vessels.
Method:	—
Summary:	Change from expensive flammable organic solvents to alkaline degreasing, from hand cleaning to staying in vessels.

EXAMPLES   143

>   **Satisfied:**   Yes, because the risks of fire + vapors of organic solvents have been removed. Gloves now fully protect.

118   CLEANING IN THE AUTOMOBILE TRADE

>   **Professional group:**   Motor mechanics.
>   **Working function:**   Degreasing of metal materials.
>   **Cause:**   Wish to avoid organic solvents.
>   **From product/process:**   Cleaning/washing off with solvents by means of a paint brush.
>   **Solution:**   Washer/water-based (basic) cleaning liquid.
>   **Method:**   Toxicity assessment, experiences of other people.
>   **Summary:**   —
>   **Satisfied:**   Yes, because cleaning is now performed mechanically (previously it was manual) and workers + management are satisfied with the quality of the cleaning.

119   OIL DIRT, CLEANING

>   **Professional group:**   Engineering industry.
>   **Working function:**   Cleaning of oil and dirt from oil grooves.
>   **Cause:**   —
>   **From product/process:**   Cleaning agent containing hydrocarbons with high-pressure flushing.
>   **Solution:**   Water-based cleaning agent with glycol + water flushing at water works pressure.
>   **Method:**   Vapor pressure comparison + emulsification property.

Summary: A hydrocarbon product with tensides (on account of the external environment) required high pressure application. A water-based product containing glycol applied with brush and flushing with water with (slightly increased) piping pressure could flush sufficiently.

Satisfied: Yes, because the quantity of aerosol decreased. No, because of difficulties in letting wastewater into the sewer arose.

120 CLEANING

Professional group: The automobile trade, motor mechanics.

Working function: Cleansing of metal surfaces.

Cause: Danger of damage from organic solvents.

From product/process: Manual use of petroleum, solvent naphta, and chlorinated organic solvents.

Solution: Use of enclosed rotating flushing plant with heated aqueous solvents.

Method: —

Summary: Reorganization of working processes from manual use of various organic solvents to use of mechanical plants with water-based cleaning agents.

Satisfied: Yes, because the organic solvents are removed.

121 CLEANING

Professional group: Iron industry, unskilled, engine fitters.

Working function: Cleaning of engine parts and tools.

Cause: Health hazard from organic solvents.

From product/process: Chlorotene and Varnolene.

EXAMPLES

Solution:	Alkaline cleaning with Snow-Clean (potassium hydroxide).
Method:	Toxicity assessment.
Summary:	—
Satisfied:	Yes, because the process is entirely enclosed, it takes place in a washing machine.

122   CLEANING LIQUID

Professional group:	Car repairs, skilled.
Working function:	Dewaxing, engine washing, and cleansing of separate engine and gearbox parts.
Cause:	Wish to use products which are least hazardous to health.
From product/process:	Non-oil (high-boiling petroleum fractions, aromatics < 0.1%).
Solution:	Alkaline product: HP 2000 from Pers Kemi.
Method:	Toxicity assessment.
Summary:	Cleaning liquid for automobile trade containing organic solvents is to a great extent replaced by alkaline cleaning liquid.
Satisfied:	Yes, because the agent is effective and less hazardous to health.

123   CLEANING OF ENGINE ROOM

Professional group:	Automobile trade.
Working function:	Cleaning of engine room in connection with check.
Cause:	Workers were encumbered by irritation of the mucous membranes.

From product/process:	Cleaning liquid consisting of one part of low aromatic petroleum and one part of petroleum with high aromatic content, and some cyclohexanol was applied and removed by spraying.
Solution:	Spraying is now performed with warm water under medium-high pressure.
Method:	Toxicity assessment from the Occupational Health Service combined with the goodwill and creativity of the workers.
Summary:	—
Satisfied:	Yes, because the workers themselves found the best solution.

124 CLEANING OF SURFACES

Professional group:	Engineering shop, unskilled.
Working function:	Cleaning of glass fiber surfaces, which are dusty and slightly sticky. Sometimes labels have to be removed from the surface.
Cause:	The workers have headaches and skin problems.
From product/process:	Methylated spirit is used.
Solution:	Cleaning agent containing anionic tensides, isopropyl alcohol, and ammonia. (Miljø Vinduesrens from A1-REN.)
Method:	An agent containing less solvent. Toxicity assessment.
Summary:	For the purpose of cleaning glass fiber surfaces methylated spirit is replaced by a cleaning agent with, among other things, 10% isopropyl alcohol.
Satisfied:	Yes, because the workers no longer have headaches and pronounced skin irritations.

EXAMPLES 147

125 CLEANING LIQUIDS

Professional group:	Chemical industry, unskilled.
Working function:	Bonding, cleaning of PVC materials.
Cause:	Vapors from organic solvents.
From product/process:	Methylene chloride—solvent adhesive.
Solution:	1,1,1-trichloroethane—solvent adhesive. In some places, alkaline cleaning liquid and adhesive based on N-methyl-pyrrolidone were used.
Method:	VHR and toxicity assessment.
Summary:	Methylene chloride replaced by 1,1,1-trichloroethane results in an improvement of VHR of factor 7 and reduction of permeability. Methylene chloride replaced by alkaline cleaning liquid.
Satisfied:	Yes, but moderately as concerns 1,1,1-trichloroethane. The use of alkaline cleaning liquid, which has been successful in some places, reduces the strain.

126 CLEANING

Professional group:	Telecommunications, typographer/printers.
Working function:	Cleaning of printing machines.
Cause:	The harmful effects of trichloroethylene.
From product/process:	Blankrola solvent with trichloroethane.
Solution:	Drubin washing agent 590–704, containing aliphatic mineral oils.
Method:	Toxicity assessment, replacement of aromatic with aliphatic substance.
Summary:	—
Satisfied:	Yes, because the result is technically equally good while it is better for the health.

127 CLEANING LIQUIDS

Professional group:	Chemical industry, the AP-trade, unskilled.
Working function:	Lamination of materials of glass fiber reinforced polyester.
Cause:	In connection with a survey of the styrene strain.
From product/process:	Diluent contains, among other things, toluene and methanol, MAL code 5-3.
Solution:	Acetone, MAL code 4-1.
Method:	MAL and toxicity assessment.
Summary:	As a secondary task in connection with the survey of the strain.
Satisfied:	Yes, because it was a success, acetone is often used in the AP-trade. No, because acetone is a tough solvent, so regarding strain on the central nervous system, the advantages of substitution are moderate.

128 CLEANING

Professional group:	Telecommunications, technicians, and electricians.
Working function:	Cleaning of various materials.
Cause:	Irritating odor of pyridine in methylated spirit. Pyridine is used as a denaturant.
From product/process:	A-spirit 93% + methylated spirit 93%.
Solution:	Isopropyl alcohol 99.9%.
Method:	Toxicity assessment.
Summary:	—
Satisfied:	Yes, because the odor became tolerable and the result was the same.

EXAMPLES 149

129 DEGREASING AGENT

Professional group: Iron and metal industry (voluntary), unskilled, and smiths.

Working function: Degreasing of materials—manually.

Cause: Complaints of slight illness—headache, etc.

From product/process: Solvent naphtha 80/110.

Solution: Odorless petroleum.

Method: Experiences—toxicity assessment marking.

Summary: Replacement of the degreasing agent solvent naphtha 80/110 by odorless petroleum. The unpleasant effect is reduced.

Satisfied: Yes, because the health of the operators improved. No, because petroleum is dangerous, too.

130 DEGREASING AGENTS

Professional group: Automobile trade, motor mechanics.

Working function: Degreasing of metal.

Cause: Description of used materials.

From product/process: Low-boiling organic solvent.

Solution: High-boiling organic solvent, for some tasks water-based.

Method: Toxicity assessment.

Summary: A description of all products used has been made together with a list of fields of use with priorities in relation to damages to the health.

Satisfied: Yes.

131 DEGREASING OF ENGINE PARTS

Professional group: Truck garage, motor mechanics.

Working function: Cleaning of engine parts.

Cause:	Order from the Danish Working Environment Service about improved exhaust.
From product/process:	Organic solvent was applied with a paintbrush and drying was carried out by means of compressed air.
Solution:	Washing in a washing machine with hot water degreaser, based on n-paraffin.
Method:	Enclosed process.
Summary:	Washing of the engine parts in white spirit was substituted by a washing machine, and the degreasing agent was changed to one based on n-paraffin.
Satisfied:	Yes, because enclosed process is better than open handling.

132   DISINFECTOR

Professional group:	The plastics industry—hospital equipment, unskilled.
Working function:	Wiping off of tables, etc.
Cause:	Instructions were made.
From product/process:	Isopropanol 35%, Rodalon (10%) 0.1%, nitrite 0.15%, distilled water 64.14%.
Solution:	Rodalon (10%) 10%, distilled water 90%.
Method:	Inquiry at DAK (Dansk Apotekerforenings Kontrollaboratorium).
Summary:	Old traditions. It is forgotten to seek other solutions before something is insisted on. The former product should be labeled as toxic and it also contained solvents.
Satisfied:	Yes, because a solvent and nitrite were removed, but to a slight degree.

EXAMPLES   151

133   LABORATORY CLEANUP

    Professional group: Medicinal industry, laboratory assistants.

    Working function: Cleaning up of utensils.

    Cause: Irritating odor and headache.

    From product/process: Acetone.

    Solution: Acetone replaced by ethanol.

    Method: —

    Summary: Ethanol has much less irritant effect than acetone.

    Satisfied: Yes, because now the employees do not experience any irritant effects.

134   SPOT-CLEANING OF TEXTILES

    Professional group: Seamstresses.

    Working function: Oil spot-cleaning by means of 1,1,1-trichloroethane.

    Cause: Inhalation of 1,1,1-trichloroethane.

    From product/process: 1,1,1-trichloroethane.

    Solution: Use of 1,1,2-trichloro-1,2,2-trifluoroethane at the same time as point-exhaust is used.

    Method: —

    Summary: —

    Satisfied: Yes, because the health hazard has been reduced. No, because the conditions of solubility are not quite as favorable.

135   ASSESSMENT OF CLEANING PROCESS

    Professional group: Engine fitters, unskilled.

    Working function: Cleaning of building and engine parts.

Cause:	Cleaning agent with solvent used in high-pressure cleaner.
From product/process:	Engine cleaning in high-pressure cleaner.
Solution:	Skipper (company J.C. Miljørens) + flushing with water by common water works pressure.
Method:	Background knowledge + odor.
Summary:	A loose construction resulted in dust and oil on the engine parts. Requirements of cleaning were reduced, which led to replacement of high-pressure cleaning with engine cleaner for common flushing with alkaline cleaner without solvent (pH = 9).
Satisfied:	Yes, because both method and product are less harmful.

136 ANTIRUST TREATMENT

Professional group:	PVC processing, sheets, foils, pipes, engine operators.
Working function:	Antirust treatment of tools for plastic processing and subsequent cleaning of mineral oil with petroleum.
Cause:	Presence of hazardous substances in the air when tools are started, and by cleansing with petroleum.
From product/process:	Mineral oil and petroleum used for, respectively, antirust treatment and cleaning.
Solution:	Antirust treatment with water-based cleaning agent containing anticorrosive substance. No need for cleaning when tools are started.
Method:	—
Summary:	Reorganization of the working process from mineral oil as rust inhibitor with subsequent cleaning with petroleum, to water-based cleaning agent with an anticorrosive substance.

Satisfied: Yes, because mineral oil and petroleum are removed from the working environment and replaced by a water-based product, which makes subsequent cleaning unnecessary.

137 RUST DISSOLVING AGENT

Professional group: Rental company, unskilled.

Working function: Maintenance of rental equipment.

Cause: Question whether the product "Bernal Super 6" is dangerous, and possible substitution.

From product/process: Bernal Super 6, which contains a solvent.

Solution: Water-based "Castrol rustilo Aqua 53."

Method: Toxicity assessment.

Summary: —

Satisfied: No, because the product was not satisfactory, as the rental equipment did not attain the intended shiny surface.

138 PORCELAIN POLISHING

Professional group: Ceramics industry—porcelain, polishers.

Working function: Polishing of glassware.

Cause: Exceeding the threshold limit value of respirable quartz.

From product/process: The abrasive consisted of a very finely ground glaze containing quartz.

Solution: After various experiments, finely ground Molochite (burnt kaolin, less than 1% quartz).

Method: —

	Summary:	Quartz and other materials which may cause pneumoconiosis are ordinarily not used in processes without warrantable ceramic reasons. Since the quartz is no constituent part of the product, it must be possible to replace it in this case.
	Satisfied:	Yes, because the new abrasive works.

139   SANDBLASTING

	Professional group:	Iron and metal, unskilled.
	Working function:	Assistant function for sandblast operator.
	Cause:	Expectations of the safety organization of high strain of alpha quartz.
	From product/process:	Sandblasting with flint as blast agent, content of crystalline/amorphous silicic acid between 70 and 100%.
	Solution:	Change of the blast agent to aluminum silicate, approximately 50% amorphous silicic acid, and iron and aluminum oxide.
	Method:	Toxicity assessment.
	Summary:	Measurements showed a considerable strain of alpha quartz with flint as blast agent. Change of blast agent and improved ventilation showed, by control measurements, an unequivocal reduction of the strain.
	Satisfied:	Yes.

140   RELEASING AGENT

	Professional group:	Cement works, unskilled.
	Working function:	Operator of block and slab machine.
	Cause:	Irritating smell and fear of permanent damage to health.

EXAMPLES 155

From product/process:	Mineral oil diluted by diesel/petroleum is used as a releasing agent in the concreting procedure of block and slabs. Applied by spraying.
Solution:	Use of water as a releasing agent.
Method:	Current norms. Trend away from products based on mineral oil.
Summary:	Unfortunate exposure to mineral oil with diesel/petroleum is removed by the use of water.
Satisfied:	Yes, because water works well as a releasing agent in this case. No, because slurry runs to the plate where the materials are concreted, and when the water dries a dust problem arises.

141 RELEASING AGENT FOR MOLDS

Professional group:	Concrete works, unskilled.
Working function:	Concrete releasing agent.
Cause:	Problems of air pollution.
From product/process:	Based on mineral oil.
Solution:	Water-based releasing agent for molds.
Method:	Toxicity assessment.
Summary:	New water-based releasing agent as replacement of the one based on mineral oil.
Satisfied:	Yes, because the new one is less hazardous to health.

142 OIL FOR BUILDING/RELEASING AGENT

Professional group:	Building contractors, concrete unit factory.
Working function:	—

156   SUBSTITUTES FOR HAZARDOUS CHEMICALS IN THE WORKPLACE

Cause:	Fear of damages from mineral oil, organic solvents.
From product/process:	Based on mineral oil, containing solvents.
Solution:	Water-based product, emulsion of synthetic ester oil (vegetable).
Method:	Toxicity assessment.
Summary:	Total recipe from the manufacturer has been assessed in the center. Test runs in a couple of companies.
Satisfied:	Yes, because mineral oil and organic solvents are avoided and because the response from the companies is positive.

143   OIL

Professional group:	Scaffolding, unskilled.
Working function:	Dust-laying agent.
Cause:	Problems of surfaces.
From product/process:	Product based on mineral oil.
Solution:	Vegetable oil containing petroleum. Water has also been tried.
Method:	Toxicity assessment + technical properties.
Summary:	Vegetable oil with petroleum is used for dust-laying of scaffolding boards.
Satisfied:	Yes, because the product is less hazardous to health and works technically. No, because water did not work technically.

144   LUBRICANT/RELEASING AGENT

Professional group:	Tileworks, unskilled.
Working function:	Molding of clay to tiles. Lubricant/releasing agent is used on knives placed near the breathing zone of workers.

EXAMPLES 157

Cause:	Headache.
From product/process:	Petroleum and train oil in the proportion 1:2.
Solution:	Ester oil.
Method:	Toxicity assessment.
Summary:	To remove petroleum, experiments were made with pure ester oil, but the supplier (KVK) did not want to supply in the small quantities in question. The workers have rejected several experiments.
Satisfied:	No, because the workers were not satisfied with the technical results. The supplier did not support the project.

145 RELEASING AGENT

Professional group:	The plastics industry, unskilled.
Working function:	Wax treatment of molds.
Cause:	Content of chlorinated solvent.
From product/process:	Agent with 46% trichloroethane, 53% special benzene without aromatics.
Solution:	Agent with 60% freon 11, 38% special benzene without aromatics.
Method:	General toxicity assessment.
Summary:	—
Satisfied:	No, but the cause is other circumstances of the use of the agent.

146 RELEASING AGENT FOR PLASTIC PARTS

Professional group:	Toy factory, unskilled.
Working function:	Joining of plastic parts.
Cause:	The releasing agent used caused irritant effects on the workers and released tension.

From product/process:	Releasing agent consisting of common rinsing agent.
Solution:	Change to releasing agent consisting of 89.6% water + 3.2% sulfatol B10 + 3.2% texapon N40 + 4.0% sodium benzoate. In the second phase change of this material, so that releasing agent was redundant.
Method:	Toxicity assessment, assistance from adviser.
Summary:	A releasing agent consisting of a common rinsing/flushing agent was replaced by an agent of water and sodium dodecyl sulfate. In the second phase, the materials were changed, so that the use of releasing agent could be avoided.
Satisfied:	Yes, because the irritant effects on the workers disappeared. Finally, the use of chemicals could be avoided.

147 RELEASING AGENT FOR WORK IN GUTTERS

Professional group:	Plumbers.
Working function:	Work in gutters.
Cause:	Eye irritation.
From product/process:	Ecological releasing agent, 10% silicone, spray (Helge Bülow).
Solution:	Soft soap.
Method:	Toxicity assessment.
Summary:	Plumbers got eye irritations from using releasing agent with silicone spray for gutters. Changed to soft soap.
Satisfied:	Yes, because everybody is satisfied.

148 PETROLEUM IN PUNCHING MASS

Professional group:	Ceramics industry, ceramic workers.
Working function:	The production and use of ceramic punching mass.
Cause:	Petroleum which was used hitherto was eliminated from the supplier's production. General wish to reduce the evaporation of aromatics in working areas.
From product/process:	Petroleum added to the punching mass in order to achieve a proper consistency during the punching and at the same time a lubricating effect in the punching work.
Solution:	Exxsol D80 from Esso, an aliphatic solvent with a low content of aromatics and a low evaporation number.
Method:	Assessment of odor in working areas and function in the punching and the subsequent process.
Summary:	Petroleum was replaced by Exxsol in a batch punching mass and the substructures thus produced were followed through the burning to the final process of sorting the burned porcelain cups.
Satisfied:	Yes, because the workers are satisfied, and no problems of production have arisen.

149 KEROBIT

Professional group:	The oil trade, refinery technicians.
Working function:	Antioxidant for visbroken naphtha.
Cause:	Some persons are extremely allergic to Kerobit.
From product/process:	Kerobit.
Solution:	Annulex.

Method:	—
Summary:	Thorough examination within the concern in order to find an alternative.
Satisfied:	Yes, because Annulex is nonallergenic.

150 COLD MOLDING COMPOUND

Professional group:	Gas and electricity, gas and electricity workers.
Working function:	Casting of a solid mass around cable joints to protect them.
Cause:	Originally a two-component epoxy molding compound was used.
From product/process:	—
Solution:	Cold molding compound with a content of bitumen, mixed and prepared in a closed system of plastic bags by addition of a mineral oil. After mixing, the product is ready for use. It is supplied by LK-NES.
Method:	—
Summary:	To avoid an epoxy product, a product consisting of bitumen and mineral oil handled in a closed system was chosen.
Satisfied:	Yes.

151 DUST IRRITATIONS DURING BRICKLAYING WORK, DRY MORTAR

Professional group:	Bricklayers.
Working function:	Mixture of mortar and scraping of joints.
Cause:	Skin irritations and order of the Danish Working Environment Service.
From product/process:	Dry mortar.

Solution:	Wet mortar or fully automated mixing plants, which are totally closed. Product development of dry mortar.
Method:	Toxicity assessment attempted through the Occupational Health Institute. Not yet finished. Cooperation with organizations in the trade.
Summary:	The use of dry mortar caused skin irritations and irritations of the respiratory passage because of the pressure of dust during the mixing process and scraping of joints. Traditional wet mortar is recommended, and if this cannot be used, automatic, closed mixing plants are recommended. Leaflet sent to all bricklayers.
Satisfied:	Yes, because wet mortar solves all mixing problems but they do not use it, though. No, because the final solution in all phases of bricklaying has not yet been found.

152 AUTOMOBILE PUTTY

Professional group:	Automobile trade.
Working function:	Filling of car bodies.
Cause:	Fear of health hazard from volatile solvent.
From product/process:	Toluene.
Solution:	Another solvent (termed faintly odorous) was proposed by the manufacturer: High-boiling aromatic hydrocarbons (among others, xylene, otherwise C_9 to C_{14} aromatics).
Method:	Modified VHR (no threshold limit values for parts of the new solvent).
Summary:	The Occupational Health Service asked a toxicologist for advice. He was not in favor of the substitution: "Toluene is known, but the harmful effects of the others are not known

(very well)." Laboratory measurements of the evaporation showed a difference of factor 4 in the fraction sums.

Satisfied: No, because the improvement shown by the VHR calculations was too small in relation to the uncertainty of the harmful effects of the new solvent.

153 CHANGE OF JOINT FILLER AFTER RENOVATION OF BUILDINGS

Professional group: Laboratory staff.

Working function: —

Cause: Irritant effects.

From product/process: Sadofoss elastic IK + polysulfide joint filler (also called outdoor/indoor joint filler).

Solution: Change to Bostik Bygsilicon 2685, which only degasses ethanol.

Method: Toxicity assessment.

Summary: Irritant effects as a consequence of bad indoor climate and stench. The joint filler was suspected and replaced, and the irritant effects disappeared.

Satisfied: Yes, because the irritant effects disappeared.

154 METHYLATED SPIRIT IN PRESSURE TEST CONTAINERS/ADJUSTMENT BATHS

Professional group: Iron and metal, unskilled.

Working function: Control/adjustment of apparatuses.

Cause: Irritating smell from turpentine in methylated spirit (ethanol).

From product/process: Ethanol denatured by turpentine.

Solution:	Experiments with ethanol denatured by 2-propanol (isopropyl alcohol).
Method:	Toxicity assessment.
Summary:	The experiment was not successful, as the operators complained (more) about the new smell.
Satisfied:	No, because the operators thought that the smell of isopropyl alcohol was worse.

155   SUSPENSION OF CORN STARCH—ISOPROPANOL/WATER

Professional group:	Carpet factory, dyers.
Working function:	Suspension of "starch for smoothing" of carpet dyes.
Cause:	Acute symptoms of poisoning.
From product/process:	Isopropyl alcohol.
Solution:	Coating of starch with borax, improvement of ventilation and suspension in water.
Method:	Toxicity assessment.
Summary:	Isopropyl alcohol was used in large quantities for the suspension of starch for smoothing carpet dyes. The starch is coated with Borax and may now be suspended in water. Isopropanol is replaced by water and the ventilation is improved.
Satisfied:	Yes, because symptoms disappeared.

156 PRESSURE MEASUREMENT

Professional group:	Telecommunications, unskilled.
Working function:	Pressure measurement of ventilation plant.
Cause:	Health irritations by filling of manometer liquid.

From product/process:	Struer's manometer liquid with dibromoethane and petroleum.
Solution:	Water as manometer liquid.
Method:	Toxicity assessment.
Summary:	As the manometer liquid on U tubes has a density of 1, the toxic product may be replaced by water.
Satisfied:	Yes, because adding manometer liquid to U tubes can be performed without the use of personal protective measures.

157 CREASE-PROOF TREATMENT OF TEXTILES WITH FORMALDEHYDE EVAPORATING SUBSTANCES

Professional group:	Textile workers, seamstresses.
Working function:	Operations of cutting, sewing, and finishing.
Cause:	Irritant effects because of the evaporation of formaldehyde.
From product/process:	Crease-proof treatment at dye-works.
Solution:	(1) Change of recipe. (2) Change of process.
Method:	—
Summary:	The changes of recipe and process have led to a reduction of the quantity of free formaldehyde.
Satisfied:	Yes, because the quantity of free formaldehyde has been reduced. No, because a certain quantity of formaldehyde is still being evaporated.

158 PACKAGING WITH PLASTIC FILM

Professional group:	Plastics, unskilled.
Working function:	Packaging of materials on pallets with plastic shrink film.

Cause:	Health hazard and irritating odor from the shrink film.
From product/process:	Heating by gas flame.
Solution:	Wrapping film without heating of the plastic.
Method:	Experiences of others and simple logic.
Summary:	Heated wrapping film causing irritating odor was changed to mechanical packaging without heating of the plastic film.
Satisfied:	Yes, because the irritant effects have been removed and the customers have acccpted the packaging in the new way.

159 WELDING OF STAINLESS STEEL

Professional group:	Metal industry, smiths.
Working function:	Welding of steel materials.
Cause:	Hazard of welding of stainless steel.
From product/process:	Welding process.
Solution:	Joint filler with polyurethane (MDI and TDI).
Method:	Replacement of process—overall assessment: toxicity, physical strain, noise, light, etc.
Summary:	Weldings are carried out for visual reasons and are insignificant as to strain and density. The working process was therefore changed to: use of a joint filler, which hardens sufficiently to resist wear of the surface, and cleaning, among other things.
Satisfied:	Yes, because the workers are happy not to have to weld. The working process is quicker and physically less straining. No, because a good deal of protective agents are still used, as it is a polyurethane joint filler.

160 HERBICIDE

Professional group:	Building contractor, bricklayers.
Working function:	Use of herbicide against moss, lichen, and algal growth on concrete and wall.
Cause:	Inquiry from the company about hazard to health from using the agent.
From product/process:	Ceresit-Anti-Mos.
Solution:	Change to an authorized herbicide.
Method:	—
Summary:	After the inquiry, the chemical control of the Danish Environmental Protection Agency prohibits sale of the product. The company and associated building contractors are informed of the prohibition.
Satisfied:	Yes, because the safety office of the company was satisfied with the answer.

161 LAWN HERBICIDES

Professional group:	Municipalities, gardeners.
Working function:	Gardeners spray lawns against dandelions and other weeds.
Cause:	The agents are carcinogenic or suspected of being so. Apart from this, they are hazardous to health.
From product/process:	Phenoxyacetic acid herbicides, e.g., 2,4-D.
Solution:	Optimal fertilization, watering, and mowing do not prevent weeds from seeding themselves, because the grass grows well. A hoe removes the worst. The citizens of the local authority have grown accustomed to accepting more weeds. In return they have lawns free of poison.

Method:	Toxicity assessment. Overall assessment. Process substitution.
Summary:	Instead of spraying against weeds in the lawns, they are treated so that the grass thrives. At the same time the citizens are influenced toward a new view of weeds: "What is nice?"
Satisfied:	Yes, because working environment and surroundings are not polluted by unnecessary chemicals.

162   WEED CONTROL OF BEDS AND SHRUBBERIES

Professional group:	Municipalities, gardeners.
Working function:	Gardeners spray against weeds, so that rose beds may have bare earth.
Cause:	Knowledge or suspicion of long-term damages.
From product/process:	Chemical herbicides.
Solution:	Other methods of weed control: treatment by flames (in rose beds), plants covering the ground; grass among bushes, creeping bushes, etc.; material for covering the earth: chips, leaves, peat.
Method:	Toxicity assessment. Substitution process.
Summary:	Weed control in beds may be practiced by a number of other methods instead of spraying. Prevention: plants covering the ground or materials covering the ground. Control of the weeds: treatment by flames.
Satisfied:	Yes, because working environment and the surroundings are not polluted by unnecessary chemicals. The structure of the soil is improved.

5.4 INDEX OF PROFESSIONAL GROUPS

Reference is made to numbers of examples in Sections 5.2 and 5.3.

Accumulator trade 090, 105
Aircraft industry 006, 010, 087, 098, 106
Automobile trade 001, 002, 005, 011, 015, 020, 021, 022, 109, 118, 120, 122, 123, 130, 131, 152

Bricklayers 005, 008, 061, 151, 160
Building contractors 008, 013, 016, 053, 057, 061, 142, 147, 151, 160

Carpet factory 155
Cement works, see Concrete works
Ceramic workers 148
Ceramics industry 018, 035, 048, 099, 108, 138, 148
Chemical industry 027, 029, 049, 052, 066, 091, 104, 115, 125, 127
Cleaning personnel 007, 011
Clothing industry, see Textile industry
Commerce 025, 026
Custom-made footwear 067

Designers 080
Display artists 025, 026
Dyers 155

Electricians 085, 128
Electronics industry 023, 039
Engine drivers 086
Engine fitters 022, 041, 047, 088, 089, 110, 121, 135
Engine operators 136
Engineering shop 004, 014, 032, 033, 041, 124

Furniture industry 060

Gardeners 161, 162
Gas and electricity workers 150
Glass works 092
Graphics trade 102, 103, 107

Iron and metal industry 030, 033, 038, 040, 041, 042, 043, 047, 093, 095, 110, 111, 112, 113, 114, 116, 119, 121, 129, 139, 159

EXAMPLES

Joiners 057

Laboratory assistants 048, 049, 050, 051, 052
Laboratory staff 133, 153
Leather industry 054, 059, 077

Mechanics 001, 002, 109, 118, 120, 122, 130, 131
Medical industry 050, 051, 070, 133
Metal industry, see Iron and metal industry
Model engineers 071
Modeler 018
Molders 108
Municipalities 024, 161, 162

Oil trade 094, 149

Painters 013, 014, 015, 016, 019, 020, 021, 022, 112
Plasterers 099
Plastics industry 009, 110, 046, 066, 072, 073, 074, 075, 089, 097, 101, 132, 136, 145, 158
Plumbers 053, 056, 069, 071, 147
Polishers 138
Porcelain painters 027, 029
Printers 126

Refinery technicians 094, 149
Rental company 137
Roads 019
Rubber factories 091

Scaffolding business 143
Seamstresses 034, 134, 157
Shipyards 003, 071
Shoemaking 067
Silk screen printing, see Graphic trade
Silk screen printers 017
Slaughterhouses and meat preparation 031, 037
Smiths 012, 031, 035, 038, 040, 042, 046, 129, 159
Stone industry 096
Sugars 082

Technicians 084, 128
Telecommunications 084, 128

Textile industry 017, 034, 078, 083, 088, 117, 154
Textile workers 028, 134, 157
Tile works 063, 064, 144
Tool maker 045
Toy factory 065, 076, 080, 081, 100, 146
Turners 040
Typographers 097, 126

Unskilled 004, 006, 008, 009, 023, 030, 031, 033, 036, 037, 042, 044, 054, 058, 059, 062, 063, 064, 065, 066, 068, 073, 074, 076, 077, 078, 079, 081, 082, 083, 087, 090, 092, 095, 096, 097, 098, 100, 101, 102, 104, 105, 106, 107, 111, 114, 115, 116, 117, 121, 124, 125, 129, 132, 135, 137, 139, 140, 141, 143, 144, 145, 146, 154, 156, 158

VVS (water/heat/sanitary installations) 055, 068

BIBLIOGRAPHY

General Literature on Substitution

Substitution/erstatning af kemiske stoffer og materialer, Vejledning, Branchesikkerhedsråd 5, nr. 2, 1985.
Substitution/erstatning, Kemiske stoffer og materialer på undervisningsinstitutioner, Vejledning, Branchesikkerhedsråd 12, 1987.
F. Sørensen and H.J. Styhr Petersen, Det kemiske arbejdsmiljø ved lodning, affedtning, blæsning, slyngrensning og højtryksrensning i Jern- og Metalindustrien, Rapport (inc. bilag 1, bilag 2). Arbejdsmiljøfondet 1985.
L. Seedorff, Kemiske arbejdsmiljøfaktorer, 3. Substitution, *Dansk Kemi,* 9, 1987.
E. Olsen and P.M. Andersen: Substitution: En måde at overholde lovens krav til luftkvaliteten, *Dansk Kemi,* 5, 1986.
G. Goldschmidt, Hvornår kommer der gang i substitutionen? *Arbejdsmiljø og samfund,* 4, 1986.
J. Hansen and O. Broberg, Substitution af kemiske stoffer og produkter, *Dansk Kemi,* 6/7, 1986.
F. Sørensen and H.J. Styhr Petersen, Substitution of dangerous chemicals. Abstract, Inst. for Kemiindustri, Danish Technical Highschool, Sept. 1986.
K. Larsen, *Subsitutionsmodel for autobranchens renseprodukter—eller hvordan udvælges det mindst farlige renseprodukt* (2 volumes), Ballerup-Herlev BST, 1986.
O. Banke, M. Søgaard Jørgensen, and A. Kamp, Det farlige arbejde, Fremad, 1979.
N. Overgaard, I teorien ganske simpelt, men alligevel indviklet, Arbejdsmiljø, 1, 1988.
L. Seedorff, *Substitution* (Totalvurderinger, kombinationseffekter, manglende data...). Arbejdsmiljøfondet, 1988.

Handbook Literature to Be Used in Toxicity Assessments

(Required by the Danish Environmental Protection Agency when information is collected for the assessment of the hazard of a chemical substance or product):

Registry of Toxic Effects of Chemical substances (RTECS), *U.S. Department of Health, Education and Welfare, Rockville, MD.
N.I. Sax, (Ed.), *Dangerous Properties of Industrial Materials,* 6th ed., Van Nostrand Reinhold, New York, 1985.
N.V. Steere, (Ed.), *CRC Handbook of Laboratory Safety, 4th Edition.* CRC Press, Boca Raton, FL, 1995.
F. Patty, (Ed.), *Industrial Hygiene and Toxicology,* 3rd ed., Volumes 1–3, Wiley-Interscience, New York, 1978, 1981–1982.
R.E. Gosselin, H.C. Hodge, R.P. Smith, and M.N. Gleason, *Clinical Toxicology of Commercial Products,* 4th ed., Williams & Wilkins, Baltimore, 1976.

W.M. Grant, *Toxicology of the Eye,* 2nd ed., Charles C Thomas, Springfield, IL, 1974.

Martindale: *The Extra Pharmacopeia,* 27nd ed., Pharmaceutical Press, London, 1978.

Merck Index, An Encyclopedia of Chemicals and Drugs, 9th ed., Merck & Co., Rahway, NJ, 1976.

NPIRI Raw Materials Data Handbook I–III. National Printing Ink Research Institute, Bethlehem, PA, 1974.

G. Hommel, *Handbuch der gefährlichen Güter,* Springer-Verlag, Berlin, 1980.

Stuurgroep-Kemikaarten, *Handling Chemicals Safely,* Amsterdam, 1980.

Kühn-Birett, *Merkblätter Befährliche Arbeitsstoffe,* Verlag Moderne Industrie, München 1976.

Kemikalier og sikkerhed: Sikkerhedsudvalget for Kemiske Industrier. Kemisk Forlag A/S, København (loose-leaf system).

REFERENCES

1. Bekendtgørelse om stoffer og materialer, Ministry of Labour order no. 540, 2 September 1982.
2. F. Sørensen and H.J. Styhr Petersen, Substitution of dangerous chemicals. Abstract, Inst. for Kemiindustri, Danmarks Tekniske Højskole, Sept. 1986.
3. J. Drachmann and P. Langaa Jensen, Vejledning i arbejdsmiljøundersøgelser på afdelings-og arbejdsplads-niveau, Danish Working Environment Fund, 1981.
4. Grænseværdier for stoffer og materialer, Danish Working Environment Service order no. 3.1.0.2, Copenhagen, April 1988.
5. B. Fallentin, Hygiejniske grænseværdier, I. *Håndbog i arbejdsmedicin.* Munksgaard, Copenhagen 1977.
6. D. Henschler, I Gesundheitsschadliche Arbeitsstoffe. Toxikologisch-arbeitsmedizinische Begrundung von MAK-werten, *Deutsche Forschungsgemeinschaft,* Part 6, Verlag Chemie, Weinheim, 1978.
7. Kemisk systematisk liste over grænseværdier. National Institute of Occupational Health, Denmark, Report no. 16, 1984.
8. Industri-chemikalien. Folder from BASF, 1983.
9. Miscellaneous data. Data confirmed experimentally by graduate engineer Erik Olsen, the National Institute of Occupational Health, Denmark.
10. Calculations made by graduate engineer Erik Olsen, the National Institute of Occupational Health, Denmark.
11. E. Olsen, P.M. Andersen: Substitution: En måde at overholde lovens krav til luftkvaliteten. *Dansk kemi* 5, 1986.
12. The Danish Working Environment Service order no. 464, 3rd August 1982: Bekendtgørelse om fastsættelse af kodenummer for produkter omfattet af arbejdstilsynets bekendtgørelse om erhvervsmæssigt malearbejde.
13. The Danish Working Environment Service orders nos. 360/1, 360/2, and 360/3 from 1982: Foranstaltninger mod sundhedsfare ved bygningsmaling/-skibsmalearbejde/malearbejde på store konstruktioner.
14. The Danish Working Environment Service order no. 463, 3rd August 1982: Bekendtgørelse om erhvervsmæssigt malearbejde.
15. A.B. Mikkelsen: Anvendelse af en simpel ekspositionsmodel, MEM 1, til vurdering af relativ risiko ved udsættelse for dampe under påføring af maleprodukt. Kemisk Institut, Århus Universitet, 1988.
16. The Danish Environmental Ministry order no. 662, 14th October 1987: Bekendtgørelse om klassificering, emballering, mærkning, salg og opbevaring af farlige kemiske stoffer og produkter.
17. The Danish Environmental Ministry order no. 724, 18th November 1987: Bekendtgørelse om klassificering, emballering og mærkning af kemiske produkter, der skal anvendes som opløsningsmidler.
18. The Danish Environmental Ministry order no. 725, 18th November 1987: Bekendtgørelse om klassificering, emballering og mærkning af farlige malevarer.
19. The Danish Environmental Ministry order no. 197, 6th April 1988: Bekendtgørelse af listen over farlige stoffer.

20. The Danish Working Environment Service order no. 181, 13th January 1988: Bekendtgørelse om materialer med indhold af flygtige stoffer herunder organiske opløsningsmidler.
21. Sigurd Mikkelsen: Hygiejniske Grænseværdier, fra: Opløsningsmidler på arbejdspladsen—En rapport fra konferencen Arbejdsmiljø og opløsningsmidler i Århus d. 19.–20. marts 1977. Samarbejdet mellem Arbejdere og akademikere.
22. Sven Ove Hansson og Eva Hellsten: *Arbejdsmiljø fra A til Ø.* Fremad, 1982.
23. IARC monographs on the evaluation of carcinogenic risks to humans. Supplement 7. 1987.
24. Foreign lists of carcinogenic substances. The National Institute of Occupational Health Denmark, report no. 14, 1984.
25. Forslag til handlingsplan for området arbejdsbetinget kræft. The National Institute of Occupational Health Denmark, June, 1987.
26. Kræft og kemiske stoffer. The directorate of the Danish Working Environment Service, report no. 7, 1981.
27. Forslag til handlingsplan for området arbejdsbetingede reproduktionsskader. The National Institute of Occupational Health Denmark, July 1987. Ulla Hass, the National Institute of Occupational Health Denmark: Personlig meddelese.
28. Fosterskader og kemiske stoffer—redegørelse fra en arbejdsgruppe. The National Food Agency of Denmark, 1986.
29. Klaus E. Andersen: "Allergitestet"—Hvad betyder det? *Nordisk Medicin,* 99,5, 1984.
30. Klaus E. Andersen: Allergitestning er et vigtigt, men også usikkert redskab. *Arbejdsmiljø* 12, 1987.
31. N.I. Sax (ed.): *Dangerous Properties of Industrial Materials, 6th Ed.* Van Nostrand Reinhold Company, New York, 1985.
32. Ministry of Labour order no. 407, 18th September 1979: Røgklassificering af svejseelektroder.
33. Jon H. Ruth: Odor thresholds and irritation levels of several chemical substances: A review. *Am. Ind. Hyg. Assoc. J.,* 47, A142, 1986.
34. G. Hommel: *Handbuch der gefährlichen Güter.* Springer-Verlag, Berlin 1980.
35. Filip Kristensen, Ole Jakobsen: Brugsanvisningers troværdighed. *Arbejdsmiljø,* 1987:9
    Lisbeth Engel Hansen: Brugsanvisningers troværdighed—igen. *Arbejsdmiljø,* 1987:12, Christian Libak Pedersen: Brugsanvisninger ofte vildledende. *Arbejdsmiljø,* 1988:4.
36. Inquiry by telephone, Kemiservice, the Danish Working Environment Service.
37. Nordiska Expertgruppen för Gränsvärdesdokumentation. a 20. Benzen. *Arbeta och Hälsa* 1981:11.
38. Ref. 31.
39. Nordiska Expertgruppen för Gränsvärdesdokumentation. 16. Hexan. *Arbeta och Hälsa* 1980:19.
40. Nordiska Expertgruppen för Gränsvärdesdokumentation. 43. Methylethylketon. *Arbeta och Hälsa* 1983:25.
41. Nordiska Expertgruppen för Gränsvärdesdokumentation. 2. Toluen. *Arbeta och Hälsa* 1979:5.

42. Nordiska Expertgruppen för Gränsvärdesdokumentation. 15. Isopropanol. *Arbeta och Hälsa* 1980:18.
43. Nordiska Expertgruppen för Gränsvärdesdokumentation. 63. Cyklohexanon och cyklopentanon. *Arbeta och Hälsa* 1985:42.
44. Sikkerhedskort for råvarer i malevare- og trykindustrien. September 1983.
45. Nordiska Expertgruppen för Gränsvärdesdokumentation. 51. Phenol. *Arbeta och Hälsa* 1984:33.
46. Nordiska Expertgruppen för Gränsvärdesdokumentation. 61. Redestilleret petroleum (fotogen). *Arbeta och Hälsa* 1985:24.
47. Nordiska Expertgruppen för Gränsvärdesdokumentation. 64. Mineralsk terpentin/lacknafta. *Arbeta och Hälsa* 1986:1.
48. Nordiska Expertgruppen för Gränsvärdesdokumentation. 14. Etylenglykol. *Arbeta och Hälsa* 1980:14.
49. Nordiska Expertgruppen för Gränsvärdesdokumentation. 44. Propylenglykol. *Arbeta och Hälsa* 1983:27.